One of the challenges of modern life is how to live in a creative and fulfilling fashion when the culture makes such strong demands for conformity and commercially defined productivity. In her book *Women and Dionysus: Appearances and Exile in History, Culture, and Myth*, Maggy Anthony addresses this problematic and its history through the lives of women who have forged unique paths to living such lives, providing inspiration for others to forge their own paths to living creative and fulfilling lives.

Oliver Ocskay, PhD Licensed Psychologist, US

T0299862

Women and Dionysus

Women and Dionysus links repression of the Dionysian spirit in Western culture with the rise of the patriarchy over the course of two millennia. It effectively draws a connection between Dionysus and women throughout history, with examples from cultures both past and present, and the author's own experiences.

Maggy Anthony explores Dionysus' role as god of the vine, creativity and passion, and his impact on art and literature. The book examines the Dionysian influence on creative older women, including Georgia O'Keeffe, Martha Graham and Marguerite Duras; examines Dionysus in mythology, history and religion; and considers connections to mysticism and the Renaissance. Anthony goes on to explore how women's expressions of creativity through healing, wine-drinking and dancing were condemned in history, and how modern African and Latin American rites contrast with Western traditions. Finally, the book looks at "outbreaks" of modern Dionysian spirit – from Haight-Ashbury to the Burning Man festival – and speculates on its future.

This unique study will be essential reading for academics and scholars of Jungian and post-Jungian studies, and for analytical and depth psychologists, particularly those with an interest in female individuation, creativity, and spirituality.

Maggy Anthony studied at the C.G. Jung Institute in Zurich, Switzerland, and the Zurich Clinic and Research Center for Jungian Psychology, and is a former family therapist at St. Mary's Medical Center/Maclean Center for Addictions and Behavioral Health in Nevada, USA. She is the author of several books, including *Salome's Embrace: The Jungian Women* (Routledge, 2018) and currently resides in Nevada.

Routledge Focus on Analytical Psychology

The Routledge Focus on Analytical Psychology series features short books covering unique, distinctive and cutting-edge topics. For a full list of titles in this series, please visit: www.routledge.com/Routledge-Focus-on-Analytical-Psychology/book-series/FOAP

Titles in the series:

A Primer of the Psychoanalytic Theory of Herbert Silberer
What Silberer Said by Charles Corliss
Jungian Theory for Storytellers: A Toolkit by Helen Bassil-Morozow

Women and Dionysus
Appearances and Exile in History, Culture, and Myth
Maggy Anthony

Women and Dionysus

Appearances and Exile in History, Culture, and Myth

Maggy Anthony

Routledge
Taylor & Francis Group

LONDON AND NEW YORK

First published 2021
by Routledge
2 Park Square, Milton Park, Abingdon, Oxon OX14 4RN

and by Routledge
52 Vanderbilt Avenue, New York, NY 10017

Routledge is an imprint of the Taylor & Francis Group, an informa business

British Library Cataloguing-in-Publication Data
A catalogue record for this book is available from the British Library

Library of Congress Cataloging-in-Publication Data
A catalog record has been requested for this book

ISBN: 978-1-138-61044-6 (hbk)
ISBN: 978-0-429-46579-6 (ebk)

Typeset in Times New Roman
by Newgen Publishing UK

This book is dedicated to Dr. Charles Ponce,
who introduced me to Dionysus in analysis,
assuring me I had met Him before,
many years ago in childhood.

And, as always, to the memory of Bettilou C. O'Leary
(1929–1972)

Saravá Iémanjá!

And lost the day to us in which a measure hath not been danced. And false be every truth which had not laughter along with it.

Friedrich Nietzsche, *Thus Spake Zarathustra*
(Thomas Common translation, Macmillan,
New York, 1911)

Contents

x   *Contents*

Acknowledgments

As always, the support and encouragement of friends and colleagues have gone into this work. First and foremost, to my dear friend and colleague, Andrea Thorsen Turman, who was there always, both in the hard times and the more flowing ones, with her intelligence and editing skills, and, most of all, her friendship. This book quite literally would not have come into being without her.

And Dr. Charles Ponce, depth psychologist and friend, who when Dionysus appeared once more in my life was able to help me understand what was wanted and why.

Dr. Oliver Ocskay, a psychologist with a Jungian orientation, whose conversations with me, sometimes in a late-night call, helped clarify the direction I wanted to take with this book.

In my now very long life there have been those always with me, encouraging me and giving me insights when my vision dimmed. One of those is my longtime friend, Stephen Kathriner, who with visits and frequent phone calls and texts kept my spirits up during the long writing of this work. To Professor Paul Bishop, whose studies and books about Jung and Nietzsche were so helpful. To John Roth, who shared his experience of listening to and seeing Jean Houston lecturing at Pacifica Graduate Institute. And my niece, Dr. Renee O'Leary, whose research skills were so helpful, as in the past.

My children, Anna Justine Krave and Joshua L. Anthony, now grown adults on their own, have been supportive through all their lives, through the good times and the difficult ones.

I feel fortunate to still have a "big brother," Lou O'Leary, who, in his nineties now, is still supportive, as is my sister-in-law, Judith O'Leary.

Mike Franks provided important books when needed, and brought over countless British mysteries to relax and distract me weekly.

xii *Acknowledgments*

Finally, to my editor at Routledge, Susannah Frearson, who has helped me through the process of two books with graciousness and understanding.

Every effort has been made to contact the copyright holders of material in this book. Any omissions brought to the attention of the author will be corrected in a future edition.

Introduction

The idea and image of Dionysus seized me when, in 1992, I was in my fifties and once more in the middle of analysis. He came at night, usually in my sleep. Not for Him the daylight of His brother, Apollo. Both Semele, the Goddess of the Moon, or earlier, Persephone, Queen of the Underworld, were said to be His mother, conceiving Him by Her father, Zeus. In the midst of sleep, I, who had never written a poem, would begin to write a line in a dream, and waking, a whole poem would emerge, insisting to be recorded. That the poems came from Dionysus was obvious from the imagery in them, and the fervor. I realized this was a new sort of creative inspiration, coming from what Garcia Lorca described as the "blood-filled room of the soul." A duende. In speaking of duende, Lorca wrote that inspiration arises from the soles of the feet, up through the gut, and out through the voice, or the paintbrush, or the pen of the artist. In my case, the pen.

I wanted to know more about Him. I found that Dionysus eludes, and delights in doing so. To speak of Him we must develop a tolerance for, if not a liking of, ambiguity. He is often seen in what is not, as opposed to what is. He is more in the shadow of things.

I began my research starting in classical Greece. I was to learn that, though honored there, He was not Greek but most probably Thracian or even Indian. One of Dionysus' titles was The Stranger, and since xenophobia was prevalent then, as now, this led Him to be regarded with ambivalence, particularly in those times of early patriarchy.

As my studies progressed, I began to realize that the exile and repression of Him was linked to that of women in Western culture – not surprising given that He was known also as the God of Women. I also saw that the many attributes assigned to Him were often seen in an ambivalent light: sexuality, trance, ecstasy, the arts, moisture, and regeneration of the earth. And among the women in his retinue were also the Muses.

I found, too, that many of the manifestations of these attributes were used as negative projections onto cultures regarded as inferior or primitive by our Western culture. This continues into the modern day, even with colonialism decades behind us, but the exception of the attitudes of that era remain.

What staggered me was to learn was that the fear and hatred of women reached back more than 2,000 years or more to the beginnings of patriarchy, and even today still takes us by surprise. Far back in classical Greece, Plutarch wrote of the near-epidemic levels of suicide of women, in a society that feared and hated them. Wolfgang Lederer (1968, p. 3) wrote: "The Greek, of course...familiar, sunny, rational and male; whereas the Goddess reigned in Knossos and Mykene, dark, ancient and uncanny."

Through those attributes, and His association with women, Dionysus is linked indelibly to the Goddess, who Herself was exiled too, more than 2,000 years ago, with the rise of the patriarchy – and even more forcibly when Christianity became the main religion of the West. We shall see, however, that She continues to emerge in different forms into modern times, as does Dionysus, Her son and champion.

Reference

Lederer, Wolfgang (1968) *The Fear of Women*. New York: Harcourt Brace Jovanovich.

1 The Dionysian body

We are Dionysian from birth. Infancy and the first few years of life are all about the body and feeding and nurturing its desires. Hunger, body discomfort, all bodily sensations cause infants and small children to call to us in their displeasure so as to ease their symptoms, whether with food for their hunger or the changing of a diaper for their comfort, or just holding and soothing. It is the body and its needs which dominate.

As time goes on, our parents and our society teach us to repress those demands on them and learn to do for ourselves what we can. Some of these teachings, given harshly or with impatience, induce us to repress those needs. When very young we are eager to please to gain approval, therefore many of us even engender a forgetfulness.

At puberty, there is a psychological call to begin to be individuals NOT dependent on family care. And our bodies begin to express new needs which can be frightening to the teen, particularly if there is no trusted adult close by with whom to speak. And then it is not easy, especially in Western society which still holds tight to not being open to sexuality – or even sensuality – except on a commercial level. The Dionysian cry for gratification of the young body's need for more adult sensual pleasures becomes louder and more demanding. And this point, without any rituals of the ancient cultures for our young to transform them into adults, is where our society is breaking down. This is when some of the more negative alternatives begin to surface: drugs, alcohol, overeating and, increasingly in the teen and pre-teen years, suicide. Statistics are showing an increase in obesity, addiction and suicide in the young.

In a major way, the body IS the unconscious, and psyche enters there to give consciousness, on an often symbolic level, on what the body is seeking. Dionysus' cry is ignored by stifling it with drugs or food.

The association of Dionysus with the body is obvious, He being the God of Physical Pleasures: wine, sex, dancing, theater. All the Muses were part of His retinue from earliest times. And we mustn't forget His other side, where early on His father, Zeus, made Him Lord of the Underworld and its dead. He is actually a God of the Life Force itself, Zoe to the Greeks, so He spans our lives from birth to death.

Because of his identification with women and with the Great Goddess Herself, our society for the last two thousand years or more has developed many taboos about the body – mainly the female body and all its physical manifestations and mysteries. We saw this in early Greece and certainly in the Middle Ages, particularly during the years of the witch hunts and pronouncements of the Roman church with its official manual, the *Malleus Maleficarum*, in which woman's "insatiable need for sexual gratification" led them to consorting with the Devil, and was thus a threat to human men and society.

In modern times we continue to be obsessed with the female body in our fashion magazines, online and in all manner of ways. Despite current efforts to present the female body in all its forms, both heavy and thin, old and young, it is obvious that the body women aim for is a young, svelte, slender and tall one. And an older body is encouraged to be younger looking by a myriad of means: surgery, hard exercise and expensive creams and makeup. However, we continue to remain mostly deaf to what our own bodies are trying to tell us. Dionysus' cry is seldom listened to or heard.

2 The approach of Dionysus to older women

Like all Greek gods and goddesses, Dionysus had a variety of titles. Some of these, I believe, give us clues to His association with older women. One such title was The Loosener. After a lifetime of responsibility and attention and obligations and demands of everyday life, older age is a time to loosen the grids. To turn our focus onto ourselves and see ourselves more clearly, to pay attention to our own needs, and to cease to take our measure by society's standards. Psychotherapist and author Nor Hall has written, "Older women are targeted by Dionysus because they are 'mature.' The word, rooted in Latin, as matures (meaning) knowing the favorable moment for dissolving boundaries or crossing sides" (Hall 1990 p. 32).

His epiphany may well be announced in older women with a desire to be rid of all the cant, the "1000 lectures," as Lorca puts it. I experienced such feelings when, as my career as a family therapist was drawing to a close, I had a new phenomenon occur in my life. Many times, as I shut the door to my apartment, leaving it empty until my return from the clinic where I worked, the thought would come: "I could just walk away now and never return." I had a momentary longing to leave behind possessions, relationships and obligations and vanish into anonymity someplace else. Then remembrance – of my grown children, friends, family and all the positive trappings of my life – would reel me back into my everyday reality and responsibilities. This fleeting longing to leave was my first intimation of the call of Dionysus.

In society, particularly American society, older women become invisible, yet Dionysus finds us to give lie to the current belief that we are useless and expendable. He comes to re-energize us, to remind us of the indestructibility at the core of us. Another of His titles is Indestructible Life.

He is also the Dweller in the Depths and He calls to us from there to inform us and bring us out of our sleep. He is the power of the life force that the Greeks called Zoe. But this life force also calls for a dying. A dying to the old, sleep-filled days of lack of change. And change is hard, very hard. When tarot cards were first designed, possibly in the Middle Ages, the card that was used to represent Change was Number 13: the Death card. This, to remind us that each change we make involves the death of the old way of doing things, the old habits and beliefs.

Our fantasies of "being in control" of our lives are swept away by the irrational, which is precisely what Dionysus represents. His brother, Apollo, Lord of Reason, is much more comfortable to be around. But it is precisely with the eruption of the irrational, the unpredictable arrival of Dionysus (He always arrives suddenly), that we are face to face with our creativity. We recognize the moment it occurs as a sudden shifting of perception, enabling us to get to the core of what we are creating, whether on stage, at our laptops, confronting the empty canvas or in a major decision in our lives. In a split second there is a clarity that resonates deep within us. A word of modern slang covers it: we feel a "rush." Jung counseled to "go where the life is for you," and surely this is what he meant. For without these sudden arrivals, we continue to sleep-walk. These descents into what is sometimes spoken of as "madness" ("that's a mad thing to do!") are the sort of madness that Plato defined as "[a] supernatural release from the conventions of life" (Plato 1981 #265 p. 54) as opposed to "related to sickness." Great art can come from these moments, and events that change lives.

It is in older age too that Dionysus reveals Himself in the body, expressing Himself in the aches and pains and infirmities of age. Our bodies demand our attention as never before, perhaps, and in the here and now. There is no far away future to be planned for "some day." We live, when older, in the time and space of Dionysus. Our psychology is rooted in our emotions and in our remembering. As Rafael Lopez-Pedraza has written, "It is a time when Dionysus is, more frequently than we think, present in the body and the emotions that arise in the face of illness and death" (Lopez-Pedraza 2000 p. 67).

Even now, in the twenty-first century, we have no psychology that addresses all the issues of old age; "Dionysus has tried to get into therapy for a long, long time as an unwanted visitor" (Hillman 1980, p. 33), The loneliness as friends grow old and die, and the invisible fear and helplessness that ensue as one's own body declines – and even worse if one's mind declines, or the minds of close friends and loved ones.

I am living now in the Dionysian reality. My body betrays me, demands my care and attention as never before. Friends die, or, as in the case of my oldest friend, lose themselves and our mutual history and memories in dementia.

One might say that old age is the final initiation into the mysteries of Dionysus and, like all initiations, requires courage and intent. Many of us find that in the arts. Euripides wrote his play of Dionysus, *The Bacchae*, when he was almost eighty years old.

I have been fortunate in my life to have many friends, and several of those friends have been men, mostly writers or artists, upon whom I've depended for good, deep talk when I've needed stimulation in my work and life. One of them was an artist himself, and a superb professor, whom I met when he was teaching at the University of Oregon in Eugene: Robert Kostka. Upon retirement from teaching, he went to live in Ashland, Oregon, a town full of artists and the home of the famed annual Ashland Shakespeare Festival. While I lived a couple of states away, I would call him once a week to catch up on what he was doing and to speak to him about my current writing project. On one occasion I called to tell him I had met a charismatic woman in the town in which I was living, who was an actor with her own theater company. I had watched her on stage and was entranced by her performance. Afterward, I spoke to her and told her I was a writer. She asked if I had ever written a play as she was always on the lookout for new work. I lied! I said I was currently working on a play. She said she would very much like to see it when it was finished.

That evening, I went back to my apartment, took the book I was working on, and began to turn it into a play. It was a book based on the end of life of one of my favorite French writers, Marguerite Duras. I entitled it *The Acolyte*. While I was telling Bob this on the phone, I mentioned the young man who had come into Duras' life when she was already old, not in good health and at a low ebb in her writing life. He said, quite casually, that seemed to happen to many creative women: to Georgia O'Keeffe and Martha Graham, that he knew of. And perhaps, he added, to others. I immediately thought of Isak Dinesen and Colette.

I went on and wrote the play, and when finished, I showed it to the actor and her husband, who was also her partner in the theater group. She loved it and promptly signed me on as Playwright in Residence. My play was performed shortly after. I was sixty-five years old. My head was swimming. And I could almost hear a laugh somewhere from Dionysus, God of the Theater.

The words of Kostka remained with me and I began to read about Graham, and about O'Keeffe, and saw what he had been speaking of. In short, both women, and Duras, had a young man enter their lives at a point when it seemed that, due to old age, and perhaps even illness, they feared their careers were at an end. The arrival, in each case, of a young man had produced a new, late period of creativity. Being mythologically minded, and having spent a study year at the Jung Institute in Zurich, along with undertaking a Jungian analysis of several years, I perceived and was convinced that it was the archetype of Dionysus, most often depicted as a young man, that appeared in their old age to revitalize them. I had felt that He saved me at sixty when I was feeling that perhaps my creative life was coming to an end. I also remembered that He had appeared at other times in my life when things seemed particularly dark. With me, He had appeared mostly in dreams, and for a period of several months had awakened me at night with a poem coming from me – I, who had never written a poem before. I want to present these women to you now.

Martha Graham

Martha Graham was one of the greatest, most innovative dancers and choreographers of the twentieth century. She began her career in the 1930s and was still dancing and performing her very original works right into the 1970s, still dancing though crippled by arthritis and alcohol.

That she was a Dionysian woman herself was apparent to those who witnessed the dance she created: "there is another kind of knowledge that wells up from under the dance floor with a darker undertone… it is…more anguished and Dionysian" (Hirsch 2002 p. 195). Her main protagonists stepped on to the stage from ancient myth: Medea, Phaedra, Jocasta, Clytemnestra; and more modern, if still mythic, figures: Mary Queen of Scots, Joan of Arc, Heloise, the Bronte sisters.

She kept notebooks that are filled with quotes from a wide variety of scholars – Jung, Freud, Cornford, Campbell – and her own thoughts and intuitions about what she read, and that led to her choreography of these tales as dance.

> People say –
> How did you begin?
> Well – that's the question
> And who knows –
> Not I –
> How does it all begin?
> I suppose it never begins, it just continues.

Then later: "But one takes up the necessity of one's heritage and in time it may become one's 'calling,' ones' 'destiny,' one's 'fame,' one's immortality" (Graham and Nancy 1973 p. xiv), showing that she was well aware of what drove her. Quite often in *The Notebooks* there are references to, quotes about and notes on Dionysus.

Finally, she completely broke down and was hospitalized. It was while she was in the hospital that a young man appeared, a fan, who when Martha rang for a nurse to take her to the bathroom and none responded quickly enough, scooped her into his arms and took her himself. He was not a dancer, but a great fan of hers and remained a central part of her life during her last years. Under his influence, she realized with some sadness that while she could no longer dance herself, she could still be the choreographer she had always been. She began to teach others in her company to perform the roles she had danced and created herself, and even new roles. She admitted that it was difficult for her to watch others dance her roles, but she went on to create even more dramatic dances. There was always the flair of the mythological about them. She remained the spokesperson she had always been for dance, and expounded her theories to the world at large in a way she had not done before. She dressed more dramatically and became a celebrity in addition to one of the greatest dancers of all time. She was awarded the Presidential Medal, and other honors followed. The young man remained at her side, encouraging her when she faltered and giving her the courage to continue, though there were the usual warnings, fears, and, perhaps, jealousies on the part of friends and associates that happen when a younger man comes into the life of an older woman – particularly much more so in our society than when a younger woman comes into the life of an older man, whereby the older man is congratulated on his "prize." For Martha, I believe her younger man was the embodiment at that time of Dionysus, coming to her when most needed and remaining until she died.

Georgia O'Keeffe

Georgia O'Keeffe was acknowledged to be a great painter early on, and in part this was due to Alfred Stieglitz, an esteemed photographer and owner of a prestigious gallery in New York City. He was the earliest champion of her work when she appeared on the scene in that city. A woman artist was not that common or well received in the early decades of the twentieth century. Once, when asked why she seldom signed her work, she replied, "Would you sign your face?" – very personal work (Udall 2000 p. 304 [resource no longer available]). When she and Stieglitz married, he gave her the financial and emotional freedom to follow her desires in painting, even when she found her greatest and most lasting inspiration to be in the desert of New Mexico. He was reluctant at the time, being her husband and wanting her nearer to him, but he had the foresight to give her the liberty she needed.

After his death, she went to New Mexico to live on her own, permanently, not far from Taos. She said once in an interview "I've been absolutely terrified every moment of my life – and I've never let it keep me from a single thing I've wanted to do" (Laing 2016). And there she lived and painted for many years, into old age, when her sight began to fail. It was at this juncture that a young man, a potter, came to live nearby. As in Martha's situation, it was not a romantic attachment, though in both cases there was a bit of flirtation on the part of the women. Rather, there was a lasting emotional support that was provided and a stimulation of the creativity that still lingered, ready to be brought forward once more. As with Graham, friends and critics were alarmed at his entering her life. In an interview many years later he spoke of the fact that people thought they had already married, perhaps in secret to avoid criticism, and he and Georgia often had a good laugh over it. She often referred to the fact that all the men artists could have younger women, but that society found it shocking that she had a younger man in hers – fifty-eight years younger.

Once living near her and seeing her daily, her young potter friend encouraged her to plunge her hands into the clay to feel and shape when she could no longer see to paint, and this succeeded in helping her to live a bit longer and more happily, when it might have happened that she would slip into depression over her lost powers. The young God, coming in disguise into the life of an old woman, an artist, and making her last years years of creative expression and flow.

Marguerite Duras

Marguerite Duras was a French writer whose works as a novelist and then a filmmaker had a great impact on France and the rest of the Western world. Born in Indochina during the French occupation, she left for Paris as a young woman, prior to the outbreak of World War II. Her books had a great impact on the French avant-garde and her friendship during the war years, whilst working in the Underground with the man who was later to become President of the Republic, François Mitterrand, raised her image in the public imagination to that of a heroine. However, as the years passed she began to drink heavily and her health deteriorated severely. Though still held in high regard, she was producing less and less writing.

During a lecture at a college in northern France, a young man was so taken with her writing, and her presence, that he read all he could of her work, saw the movies she produced and wrote, and sent long letters to her. Finally, one evening he made a call to where she was staying in the South of France, struggling to write and live. As she was later to say, she made the "mistake" of inviting him to come to where she was staying in a hotel. This "mistake" undoubtedly saved her life. As tempestuous as their relationship became, her creativity soared and he remained at her side throughout her serious illness and rages against him. "An often silent presence, he accepted the blows, the insults, Marguerite's nastiness – oh, why am I so nasty she'd sometimes ask him quite distraught" (Adler 2000 p. 329).

It was with him at her side that she saw her autobiographical novel *The Lover* made into a highly successful film. Then, annoyed by the film, she wrote *The North China Lover* to contradict some of what was put into the film. She also wrote the story of her relationship with the young man and entitled it *Yann Andreas Steiner*, though that was not his real name: "She took away his name, his nights, his time, his loves. Yann became Marguerite's chauffeur, confidante" (Adler 2000 p. 392). There was little outrage at the difference in ages, perhaps because the French have a different view of such things. And, perhaps because of Duras' volatile temperament, her friends and family appreciated his steadfastness in the face of the difficulties. Finally, in her last days he took down her words as she lay dying, and it was titled *C'est Tout* (in English, *No More*). It is dedicated to him and to her love for him. Dionysus came through once more.

So who is this God? A miracle worker? From where did He come, and why does He seem to keep showing up? A part of the answer may come from His origins a few thousand years past. Nietzsche claimed to be "the last disciple of the philosopher Dionysus" (cited in Bishop and Gardner 2018 p. 81), But it is good to remember the words of Otto: "Let no man arrogate to himself the right to say that a god has died until the echo which remains after the departure of the last of his worshippers has been dispelled" (Otto 1989 p. xxi).

References

Adler, Laure (2000) *Marguerite Duras: A Life*. Chicago: University of Chicago Press.

Bishop, Paul and Gardner, Leslie (2018) *The Ecstatic and the Archaic: An Analytical Psychology Inquiry*. London: Routledge.

Graham, Martha and Nancy Wilson (1973) Foreword. In *The Notebooks of Martha Graham*. New York: Harcourt Brace Jovanovich.

Hall, Nor (1990) *Those Women*. Dallas: Spring Publications.

Hillman, James, ed. (1980) *Facing the Gods*. Dallas: Spring Publications.

Hirsch, Edward (2002) *The Demon and the Angel*. New York: Harcourt, Inc.

Laing, Olivia (2016) "The Wild Beauty of Georgia O'Keeffe". *The Guardian*, July 1: www.theguardian.com/artanddesign/2016/jul/01/georgia-okeeffe-tate-modern-exhibition-wild-beauty.

Lopez-Pedraza, Rafael (2000) *Dionysus in Exile*. Wilmette, Ill: Chiron Publications.

Otto, Walter (1989) *Dionysus, Myth and Cult*. Dallas: Spring Publications.

Plato (1981) *Phaedrus* (Helmhold, W. C. and Rabinowitz, W. G., trans). New York: Bobbs-Merrill.

3   Dionysus in mythology

"Myth is what we call other people's religion," Joseph Campbell once said at a conference I attended. What to us might seem amusing stories were once the living spiritual beliefs of ancient peoples. We must not forget. And as Heinrich Zimmer wrote, "in dealing with symbols and myths from far away, we are really conversing with ourselves...the mythical tradition provides us with a sort of map for exploring and ascertaining contents of our own inner life" (Zimmer 1971 p. 310).

We identify Dionysus with Greece, yet there are those who maintain he came from elsewhere. One of His titles was The Stranger, and various scholars have Him originating in Thrace, roughly parts of modern Bulgaria, northeast Greece and Turkey, or Phrygia (Anatolia) or Central Asia. Jane Harrison refers to Him as "an immigrant god" (Harrison 1992 p. 364). " Today it is not questioned that Dionysos, far from being a recent acquisition by the Greeks, is one of the oldest gods" (Isler-Kerényi 2014, p. 5). Wherever His origin, the Greeks accepted Him possibly as the most recent personification of an older religious tradition. At Delphi, home of Apollo and the Delphic Oracle, famed throughout the ancient world, there are indications that it was originally associated with Dionysus. There is even an old tale that the place where the Pythia prophesized also was the tomb of Dionysus, and the omphalos his tombstone. This seems appropriate and coincides with lines from *The Bacchae* of Euripides, where Tiresias proclaims:

> this god is a prophet, the Bacchic ecstasy
> And frenzy hold a strong prophetic element
> When He fills irresistibly a human body
> He gives those possessed power to foretell the future.
>
> (Villacott 1973 lines 298–301)

It is in Greece too where we hear first the stories of the women in His retinue. In addition to those called His nurses, and the Muses, there were those called the maenads. There are wild tales of their activities, including the eating of raw meat in ritual and even tales of them tearing men apart in their frenzies. How much of this is true, taking into account a culture where women had virtually no rights or liberties, a culture where there is a fear of women, is something we need to bear in mind. In the main, accounts from the times speak of a group of maenads calling to the baby Dionysus to awaken and join them as they run through a mountainside covered in trees and streams. According to Kerényi, these groups of women ran up Mt. Parnassos, randomly, winter included, of course in trance as it is hard to imagine they would attempt it in normal consciousness or been able to do it without injury to themselves, doing it at night. And in this running, the God grew from childhood to manhood amongst them.

Kerényi speculates this manic running "was dedicated both consciously and unconsciously to the service of Zoe; psychosomatic energies were summoned up from the depths and discharged in a physical cult of life" (Kerényi 1976 p. 218–9). The inner feelings of the women taking part were never made public, never appeared in the literature of the times. But of course there is a paucity of female writing or artistry from that time – which does not necessarily mean that there was none. What there was would have been disregarded, if recognized at all.

As is often the case in matters both ancient and modern concerning women, the reader must take into account the prevailing misogyny. Much has been made of the frenzy of the entranced maenads. Yet even some ancient male writers got beyond this at times: "In the *Bacchae* of Euripides, the modesty of the women in ecstasy is explicitly emphasized in the face of malicious stories told about them" (Kerényi 1976 p. 178).

In the dual fear of woman and the equal fear of lack of control, the idea of women experiencing ecstasy in the worship of Dionysus was a frightening combination for the men of the tightly constructed patriarchy of Athens, and later of Rome. Yet, as Dionysus was also the God of Indestructible Life, a purveyor of the spirit of Zoe, much of the ecstatic state experienced in His worship must be seen to be of a life-affirming nature. As Nietzsche wrote, "The fundamental fact of the Hellenic instinct – it's 'will to life' – expresses itself only in the Dionysian mysteries, in the psychology of the Dionysian state. What did the Hellenes guarantee themselves with these mysteries? Eternal life…" (cited in Bishop 2018 p. 81).

Of course, His Underworld connection with Death meant that this too was part of the experience. Death was most commonly associated with the shedding of the more comfortable living within accepted standards of an organized life. And this was what was most feared by the classical civilizations of both Greece and Rome, being highly structure societies. The fear was real, as "The Archetypal World of the feminine knows nothing of the laws and regulations which govern human society…it is a world which conforms completely to nature" (Otto 1965 p. 179).

The nature of the Dionysian trances can most certainly be linked to the freeing of the spirit of the maenads, who by day led structured – one might say, pre-ordained – roles in a culture which had definite ideas about the place of women. One example of this can be found in the images on a fifth-century BCE water jug:

> On the water jug we see the perfect Athenian woman. She is not poor. She is sitting down and being handed her baby by a slave or servant girl. At her feet she has her wool basket. That about sums up the answer to the question: what were the wives of Athenian citizens for? They were for making babies and making wool.
>
> (Beard 2018 pp. 36–37)

"It seems clear that one of the main results of Dionysiac possession enabled Greek women at least to defy their normal roles and participate in activities which were normally not permitted them" (Kraemer 1979 pp. 79–80); and, since these were religious trances and possessions, they were not punishable by society.

"For the Greeks ekstasis meant the flight of the soul from the body" (Wasson et al. 1978 p. 31). It is interesting to speculate on the possibility of the ecstatic experience being linked to the use of hallucinogens, as suggested in the late 1970s in a book by R. Gordon Wasson and others, *The Road to Eleusis*. In this it was suggested that the ritual wine drink given to initiates, called kykeon, was wine augmented by ergot producing effects similar to lysergic acid (LSD-25). Even more interesting is the fact that the book, coming out a few years after the social revolutions of the late sixties, which included a resurgence of feminism, the Black Power movement and others, was condemned and ignored by both the academic and scientific communities, and promptly went out of print in English. This despite the credentials of its three authors: Albert Hoffman, Swiss scientist and first to synthesize and learn the psychedelic effects of LSD; R. Gordon Wasson, ethnomycologist and author; and Carl A.P. Ruck, Professor of Classical Studies at Boston University.

Only recently has a 30th anniversary edition been published. It is my intuition that the maenads, being devotees of Dionysus, the God of the

Vine, may well have imbibed a similar mixture in their rituals. Using scientific testing from cups found at Eleusis, it has been at least provisionally accepted, as in traces of kykeon found, the truth of the Wasson book is indicated.

As mentioned in the Introduction, this was a God regarded with major ambivalence by the Greeks, and later the Romans. He was a symbol of excess, of ecstasy, of the madness that comes with worship and with creativity at times. As Edinger recognized, "He is connected to rapture, to the release of everything that has been locked up, to the blaze of life, but also to suffering, to madness and to death" (Edinger 1994 p. 144). And the fear of that which has been locked up for long remains with us today, especially in regard to women, as evidenced by the attempted obliteration of the power of the Goddess-centered religions of the most ancient times.

Dionysus was central to one of the most powerful of the mystery religions of the time, the Eleusinian Mysteries, as well as His own cult. This religion spread throughout the ancient world and many of the most famous Greeks and Romans of the time took part in its rituals, which were devoted to the story of the Goddesses Demeter and Persephone, called the Kore. The rituals were completely secret. To this day, we can only guess at what actually took place. However, Initiates of the day left scattered statements that intrigue us. Homer, in his Hymn to Demeter:

> Blessed is he among men on earth who has beheld this. Never will he who has not been initiated into these ceremonies, who has had no part in them, share in such things. He will be as a dead man in sultry darkness.
>
> (Kerényi 1967 p. 14)

And Pindar, "Blessed is he who, after beholding this, enters upon the way of the earth: he knows the end of life and its beginning..." (Kerényi 1967 p. 15). And again, Cicero: "We have been given reason not only to live in joy but also to die with better hope" (Kerényi 1967 p. 14).

What strikes me is that all these quotes from famous men of the time, when women were held in something near contempt, were speaking of a wholly feminine mystery religion, based on a story from a time before the Goddess had been brought down and stripped of Her old power. That power had obviously simply gone underground...literally, in the case of Eleusis, where the mysteries were enacted.

Three deities formed the triad at the heart of the rituals: Demeter, Kore and Dionysus. In the very last of the rituals enacted, it was a priest wearing the mask of Dionysus who presided. He was one of only two males in the enactments, the other being Hermes. "It is at once a cardinal point and a primary note in the mythology of Dionysus that he is son of his mother" (Harrison1992, p. 402).

There were, of course, other celebrated female deities in Greece. There was Athena, for whom an enormous edifice was built near Athens, on the Acropolis: the Parthenon, with an enormous statue of the Goddess Athena as the central point within it. However, She and the other female deities of that time show the hallmarks of a patriarchal remaking of the feminine. None of them – Artemis, Hera, Athena or others – are shown with rich bodies as were the statues of the most ancient Goddesses, whose breasts and even genitals were displayed prominently. The Goddesses of the patriarchal classical era in Greece are very much "Daddy's girls." Athena sprung fully formed from the head of Zeus and is usually pictured armed like a warrior; Artemis was very much the athletic twin of Apollo; and Hera was the jealous wife of Zeus, who tended to lead human females who were targeted sexually by her husband into fatal situations. One story has her telling Semele to demand to see Zeus, by whom she was pregnant, to show Himself in all His splendor, and Semele was killed by the sight for her trouble. His thunder and lightning did the job and He placed the unborn infant, Dionysus, in His thigh to await birth. Of course, there are indications in pre-patriarchy that Semele Herself was a Goddess of Lightning and Thunder. All these indicate

> a process of revision that ran through many if not all of the Greek myths, where the original goddess-oriented cultures have been modified or inverted to create a new kind of significance more in accord with the god-oriented culture of the Greeks.
>
> (Baring and Cashford 1991, p. 139)

As with much of Greek culture, these mythologies soon spread north, to Rome. The Romans, equally treating women with fear and anger, were quick to impose as much law against the Dionysian revels as they could.

> In 186 BCE, the City of Rome – THE City – makes the apparent discovery that the Bacchanalia, which happen at night, in reality provide a cover for rapes, murders and orgies. False testimony, false accusations and all the machinery of witch persecution are unleashed against Bacchanalian trances. Where does this fury against a Greek sect come from?
>
> (Clement 2014 pp. 83–4)

Clement goes on to write of the Roman law that women weren't to drink wine because it supposedly induced abortion, and that the Bacchantes were the only sect that accepted all classes of society, even slaves, to their ranks. The paranoia escalated to the point where this cult of Dionysus had to be regulated by the Praetor and the Roman Senate. In the second century CE Juvenal wrote of the Dionysian devotees in a poem,

> They're females without veneer, and around the ritual den, Rings a cry from every corner: We're ready! Bring in the men! And if the stud is sleeping, the young man's ordered to wrap Himself in a robe and hurry over. If he is not on tap, a raid is made on the slaves;... if they can't find a man, to save the day they'll get a donkey to straddle their itchy behinds. O, would that our ancient rites, at least in public shrines, were purged of these filthy acts.
>
> (as quoted in Kraemer 1988 p. 28)

If any of the regulations were broken, the death penalty was risked. Women and Dionysus were obviously a force to be reckoned with, in the eyes of classical Roman and Greek civilization.

However, it is to the Romans, or at least a Roman family living in Pompeii, to whom we owe a debt of gratitude. And to Mt. Vesuvius, for its massive eruption in 79 CE, whose lava flow perfectly preserved what has come to be called the Villa of the Mysteries. This is a Roman villa that contains one large room whose walls are covered with brilliantly colored frescoes, mostly in reds. The frescoes unfold to tell a story which is the passage of woman through the rites of Dionysus. It seems to have been used as an initiation chamber. Linda Fierz-David, a Jungian analyst, visited it in the 1950s and wrote a series of lectures she presented at the C.G. Jung Institute in Zurich. After her death in 1955, her sons presented the manuscript to Spring Books, who published it under the title *Women's Dionysian Initiation: The Villa of Mysteries in Pompeii*. It has undergone many subsequent editions. In the early twenty-first century, UNESCO provided a large amount of the finance to have the frescoes restored, and they are quite beautiful. It is the only known surviving fresco of the times which alludes to the mysteries of the Dionysian cult. "Every cult was a mystery, because the drama of the cult was a deep secret that could not be exploited, as the Greeks also said, the mystery is an arrheton...something ineffable. The experience could not be expressed in words" (Fierz-David 1988 p. 33). A modern Jungian depth psychologist, David M. Odorisio, wrote after visiting the Villa himself:

These ancient frescoes emerge from a culture and mythology entirely different and foreign from our own yet they depict with stunning psychological accuracy the transformations of conscious-ness undergone by one who is courageous enough to respond and surrender to the journey within.

(Odorisio 2015)

A word needs to be said about the Orphic religion and mysteries. Orpheus has been linked with Dionysus repeatedly: "clearly linked though they are, the most superficial survey reveals differences so striking as to amount to a spiritual antagonism" (Harrison 1992 p. 455). It appears that Orphism began as a reform of the Dionysian mysteries, making them a good deal milder and more acceptable to the society in which they developed. The maenads and Muses and nurses were left behind. Dionysus was associated with death, with ecstasy (especially with His women followers) and drunkenness, as well as with indestruct-ible life. Too heady a mixture for the Orphics who chose a path of vege-tarianism and near celibacy, all in an attempt of a cleansing of the soul. A watered down, more acceptable Dionysus.

It was, in the end, the Romans who put an end to what became called paganism – the religion of the common people. Beginning with the Emperor Constantine the Great, the old religion and its temples began to be destroyed. This continued in varying degrees with his successors until Christianity became the official religion by about 491 CE.

But if officially paganism was routed, as we shall see in the following chapters, in actuality it went underground, and its repression gave it strength when it surfaced throughout the ages into modern times.

References

Baring, Anne and Cashford, Jules (1991) *The Myth of the Goddess: Evolution of an Image*. London: Penguin Books.

Beard, Mary (2018) *How Do We Look? The Eye of Faith*. London: Profile Books.

Bishop, Paul and Gardner, Leslie, eds. (2018) *The Ecstatic and the Archaic*. London: Routledge Books.

Clement, Catherine (2014) *The Call of the Trance*. London: Seagull Books.

Edinger, Edward F. (1994) *The Eternal Drama*. Boston: Shambala.

Fierz-David, Linda (1988) *Women's Dionysian Initiation*. Dallas: Spring Publications.

Harrison, Jane (1992) *Prolegomena to the Study of Greek Religion*. Princeton, NJ: Princeton University Press.

Isler-Kerényi, Cornelia 2014 *Dionysos in Classical Athens: An Understanding through Images*. Boston, USA and Leiden, Netherlands: Brill Pub.

Kerényi, Carl (1967) *Eleusis: Archetypal Image of Mother and Daughter*. Princeton: Bollingen.

Kerényi, Carl (1976) *Dionysos: Archetype of Indestructible Life*. Princeton, NJ: Princeton University Press.

Kraemer, Ross S. (1979) Ecstasy and Possession: The Attraction of Women to the Cult of Dionysus. *The Harvard Theological Review*, vol.72, no. 1/2, January 1, pp. 55–80.

Kraemer, Ross S. (1988) *Maenads Martyrs Matrons Monastics*. Philadelphia: Fortress Press.

Odorisio, David (January 12, 2015) Touching Ecstasy: Dionysian Initiation Rites at Pompeii. ahomeforthesoul.com

Otto, Walter (1965) *Dionysus: Myth and Cult*. Dallas: Spring Publications.

Villacott, Philip, trans. (1973) *The Bacchae and Other Plays*. New York: Penguin Books.

Wasson, R. Gordon, Hofmann, Albert and Ruck, Carl A.P. (1978) *The Road to Eleusis: Unveiling the Secret of the Mysteries*. New York: Harcourt, Brace, Jovanovich.

Zimmer, Heinrich (1971) *The King and the Corpse*. Princeton, NJ: Princeton University Press.

4 Carnival as temenos

Religion in ancient times made space within its confines for the expression of what Jung called the shadow archetype. In pre-Christian eras, divinities behaved occasionally in an "uncivilized" manner. In the Greek pantheon there was Pan, Silenus, Hecate and, of course, Dionysus and his followers, who acted darkly. All of them acted with impropriety at times and were not thought of any the less for it. Christianity, on the other hand, makes no space for the shadow other than with Satan, the ultimate evil one.

Animal nature not being recognized as inevitable with the Church, with its greater focus on the spiritual world, the rational world and physical nature was regarded as inferior, with an extreme concurrent distaste for the physical body. The body that is Dionysus. All this led to the mind/body/spirit split and the repression of sexuality, passion and aggression. Aggression as sanctioned by the Church and State always being the exception. All of these animal/body traits have been repressed and sent to the nether world of shadow and denied expression. Dance is still prohibited in some Christian religions. Of course. Animals dance.

Yet there is also an underground tradition in Western Christianity which admits to a sacred knowledge within the shadow. We find hints of this in the legend that Satan was once the favorite of God, even His first son. His name was Lucifer, the Light-Bearer, the highest of all the archangels before his fall to earth. In that fall, esoteric story has it that an emerald fell from his crown and became the Smaragdine Tablet, on which was engraved "As above, so below." This tablet is also told to be the headstone on the grave of Hermes Trismegistus, the legendary ancient philosopher. This intuition of the potential incredible value of aspects of the shadow archetype is an expression of the knowledge that much energy is constellated around it.

Carnival, all over the world and across the centuries, is the time and place of the outward expression of the contents of the psyche otherwise

repressed the rest of the year. In Jungian psychology these contents are
defined as the shadow, both personal and collective. Jung spoke to this
several times: "the shadow is a living part of the personality and therefore wants to live with it in some form" (Jung 1968 p. 31). Much of what
has been historically repressed in Western European culture is what we
think of as Dionysian. And a great deal of it is held in the collective
shadow, which in many countries has its outlet in the celebration of
carnival.

One of the principal elements in carnival in all countries, culmin-
ating in massive, seemingly endless parades, is the wearing of animal
costumes and masks. I believe this to be in remembrance of the animal
ancestor itself, our own animal nature. In most sculptural and painted
representations of Dionysus, He is accompanied by animals and/or
animal--human hybrids such as satyrs. In the donning of animal masks
and costumes, the "skin" of animals, we are expressing our unconscious
awareness of our basic animal nature. We are permitting these impulses,
so successfully hidden and forbidden in rational society the rest of the
year, to be given expression in the sanctioned, sacred *temenos* of the
public celebration. Jung spoke of this in his writings:

> We are no longer aware that in carnival customs...there are remnants
> of a collective shadow figure which prove that the personal shadow
> is in part descended from a numinous collective figure. This col-
> lective figure gradually breaks up under the impact of civilization,
> leaving traces in folklore which are difficult to recognize.
>
> (Jung 1968 p. 50)

It is self-evident that the "collective shadow figure" of which he writes is
the animal ancestor itself. Our own animal nature.

> ...remember that the purpose of the Dionysian mysteries was to
> bring people back to the animal, not to what we commonly under-
> stand by that word, but the animal within...This is the experience
> everyone should have in order to find again the connection with the
> nature within, one's own nature and with the god of the primitives.
>
> (Douglas 1997 p. 80)

From another point of view, Paul Radin, cultural anthropologist and
folklorist, writes, "Viewed psychologically, it might be contended that
the history of civilization is largely the account of the attempts of man
to forget his transformation from an animal to a human being" (Radin
1953 p. 18). Carnival, even in its present-day, greatly impoverished form,

still provides a way to ritualize and contain the tremendous energy of the animal nature within. Civilization increasingly demands more of a separation from our instinctual natures and encourages conformity to a code of "civilized behavior." This always results in a repression not only of act, but of feeling. With carnival we can safely, with Church/State sanction, rebel against codes of conduct involving the expression of our sexuality and uninhibited behavior.

In carnivals past the animal costume was more prevalent than it is today, though these costumes are still widely seen. They are a holdover from more ancient festivals of a pagan religious nature when people wished to once again live out their animal nature, if only for a moment. Carnival is one of our last remnants of this deliberate, if presently unconscious, mode of celebrating it. It is as if some people realize that we must do this in order to live in our everyday world in a fuller way. Psychologist Bruno Bettelheim wrote "Only when animal nature has been befriended, recognized as important, and brought into accord with ego and superego does it lend its power to the total personality" (Bettelheim 1978 p. 102).

Another element in carnival that is almost universal and points once again to Dionysus is role reversal. It consists mostly of wearing the clothes of the opposite sex. As widespread as this is in carnival, there seems to be some "magical" element in it, even if largely unconscious. According to Hans Duerr, German anthropologist,

> Such a ritual is not only an expression for the initiates having abandoned the "world of women" and becoming members of the "world of men." It also signifies the dissolution of the separation between male and female modes of behavior.
>
> (Duerr 1985 p. 56)

Dionysus, of course, was the God of Women, and also His sexuality was ambiguous. And perhaps this "cross-dressing" is owed to deeper roots. "The priest and shaman all over the world, have been wearing woman's garb and bedecked themselves with intricate ritual ornament – including woman's own necklace – to be more effective in their metaphysical efforts" (Lederer 1968 p.153). Woman's ancient association with magic, making her both attractive and at times, feared.

This role reversal was a part of a larger reversal in the order of things. In ancient times and festivals, quite often the slaves became the masters and the masters the slaves for this day and for the "time between times." We see this in the medieval Feast of Fools, where there was a Fool King and a Fool Pope. Also, there are echoes in this from a much earlier time when a prisoner or slave was chosen to be king, then sacrificed in the king's place after a short, celebratory reign. This was widespread in cultures as diverse as the Ashanti in Africa and the ancient Greeks. "The King must die," but it was thought better to have someone to die in his place. These instances of role reversal might well be a way of shaking the grids of our perceptions to see beyond the world of appearances into what an Indian shaman in Mexico named the nagual: "The nagual is the part of us for which there is no description – no words, no names, no feelings, no knowledge" (Castaneda 1974).

I have had the privilege of witnessing carnival in two different cultures – Switzerland and Brazil – and, as one can imagine, the expressions are entirely different. In the Swiss celebration, specifically in German-speaking Zurich and Basel, where noise, bright colors in dress or vulgarity of behavior are anathema in everyday life, in the carnival (or *fastnacht*, as it is known there) the celebration is noisy, loud and often quite vulgar.

While watching the parade with my children many years ago, one of the marching men, dressed as a woman in a vulgar costume, left formation, ran over to me, grabbed me, tried to kiss me, then dragged me toward a bridge over the river. Fortunately, the children's father grabbed hold of him and pushed him away, whereupon he rejoined his contingent of similarly vulgarly dressed men.

In Brazil, where the African and Latin roots are strong, noise and color are part of everyday life, and the accent instead is on elegance and opulence, denied to 95 percent of the population the rest of the year because of crushing poverty. The music, in contrast to the *guggenmusik* of the Swiss holiday, which is purposely loud and discordant, is melodious and composed with great care and competiveness. There is a prize offered for the best song for carnival.

Also in Brazilian carnival, linking it even more so with Dionysus, the component of death is present. This appears in the form of skeleton costumes, death masks, black-clad men carrying coffins and suchlike. A great example of this was captured in the popular and beautiful Brazilian film *Black Orpheus*, filmed by the French director Marcel Camus. This was a film of the early 1960s, and it absolutely captured the spirit of the carnival as well as its mythological undertones. Both of the carnivals I attended, in 1963 and 1964, attested to this. In another form of role reversal, the poorest *favelados* or shantytown dwellers in Rio de Janeiro walk and dance around in lavish costumes of the eighteenth century in Brazil, when a Portuguese emperor reigned and the Black people there were slaves. Antonio Carlos Jobim, a friend from that time and one of Brazil's greatest composers, wrote a song, "Sadness Has No

End," for *Black Orpheus*, one of the lines of which, written by poet Vinicius de Moraes, says "the people work all year, for one moment of a dream, to become a fantasy of a king..." (Jobim 1961). Beautiful music, in great contrast to the ugly, purposely loud and discordant music of Swiss carnival. Yet both carnivals serve to allow a one-time expression of desires and behaviors not permitted the rest of the year.

And so, in some countries we still have sanctioned access, once a year, to our Dionysian impulses. It is, however, not quite enough.

References

Bettelheim, Bruno (1978) *The Uses of Enchantment*. New York: Knopf.

Castaneda, Carlos (1974) *Tales of Power*. New York: Simon & Shuster.

Douglas, Claire (1997) *Translate this Darkness*. New Jersey: Princeton University Press.

Duerr, Hans Peter (1985) *Dreamtime: Concerning the Boundary between Wilderness and Civilization*. Oxford: Oxford University Press.

Jobim, Antonio Carlos (1961) *Black Orpheus* [film]: soundtrack.

Jung, Carl Gustav (1968) *Archetypes of the Collective Unconscious*, Vol. 9. Collected Works. New Jersey: Princeton University Press.

Lederer, Wolfgang (1968) *The Fear of Women*. New York: Harcourt.

Radin, Paul (1953) *World of Primitive Man*. New York: H. Shuman.

5   Dionysus in the Early Middle Ages

> The present adherents of the Judaic monotheistic and…their post-Christian substitutes are, all of them ex-pantheists.
>
> (Baring and Cashford 1991 p. 298)

The evidence of this lies in plain sight in the Gothic cathedrals spread all over Europe. These magnificent buildings, which began to be constructed shortly after Pope Urban II called for the First Crusade to begin in 1095, are storehouses of the pagan roots of Christianity.

On my first trip to Europe in 1960, a close friend of mine, Brother Antoninus, a Dominican monk and poet, told me to visit as many cathedrals as possible. At the time this was not pleasing to my then agnostic ears. Not wanting to visit any "churches," however, I took his advice. In Paris, I went to Saint-Julien-le-Pauvre and was impressed by its beauty. As I exited by a side door, I was confronted with an archaeological dig in progress; I saw that directly under the church was a temple dedicated to a Roman Goddess, and, under that, an older, very primitive but obvious place of worship. As Antoninus had told me, each cathedral or church built in the Middle Ages had been built on the site of a pagan temple, and that upon an even earlier holy site.

As I traveled on, I visited mainly Gothic cathedrals and was surprised to see carvings of what I was later to learn were called the Green Man. Even, occasionally, rather "rude" carving of women gleefully exposing their genitals. This all rather confused me as, from the little reading I had done by that time, I was aware that the Church (meaning Rome and the Vatican) had specifically wiped out all traces of pagan statuary and temples. So, to find what were obviously pagan images in the most revered of medieval cathedrals came as a surprise, to say the least. I was to learn that wiping out stone buildings and temples is a great deal easier than wiping out belief in people's hearts, and obviously the Church learned this early on.

Early on, too, the Church found that a substitute, a sanctioned one, was needed for the all-powerful Great Goddess, whose worship had only been partially eradicated by the patriarchy preceding the foundation of Christianity as the religion of Western Europe. Slowly, skillfully, the figure of Mary, the mother of Jesus, was brought into focus, often with many of the trappings of the Mother Goddess who preceded her.

The building of the Gothic cathedrals was begun initially to provide a space for all the relics being brought back by the first Crusaders. As the spoils of war, they were bringing back innumerable works of art. Many of these were pagan, but readily adapted to Christian imagery when imposed upon them. The ideas behind them were archetypal after all. What was less apparent, perhaps, were the ideas that so many of the returning Crusaders brought with them, getting a broader education from their travels to lands only heard about previously from the pulpits of the churches:

> The Crusades tore down the barrier between the erudite world of Islam and the ignorant one of Christendom. Few in Western Europe knew of the incredibly rich legacy of the classical world. With an ironic twist, fate had entrusted Islam, the West's sworn enemy, with its birth records.

> (Shlain 1998 p. 293)

With all the spoils of war being brought back from the Holy Land by the Crusaders, the Green Man comes into focus, he who began to appear in carvings in the newly built cathedrals. Carvings of him are always surrounded by leaves; leaves that emerge from his mouth at times, that crown him. One of the epithets of Dionysus comes to mind: Indestructible Life. Life so abundant that it spills forth and surrounds this God. Here He is once more, reborn in the twelfth century and unrecognized. Except, perhaps, by the masons and carvers of those great cathedrals. Had they heard of Dionysus from the Crusaders? Equally possible is that in addition to those returning warriors, they heard from their colleagues in Spain too, the Arabs who had passed along their knowledge and ideas, contributing to many architectural features of those buildings, particularly the Pointed Arch. For as erudite as the Islamic culture was in the Middle East, so too was that culture in Spain and Northern France, still flourishing in the early Middle Ages. Indestructible Life. The Green Man. But Islam had their own Green Man: Khidr.

"Khidr is the personal guide…equivalent to…the Hidden Iman, to the Christ of the Cross or Light…And he…is the 'Verdant One.' He is the Green Man. He is the Angel of the Earth" (Cheetham 2005 p. 122). Khidr appeared in a vision to one of the premier Sufi mystics and teachers, Ibn Arabi, a native of Spain whose life spanned the late eleventh through the mid-twelfth centuries – the time of the Gothic cathedrals and the Crusades. So not only were these Christian cathedrals invaded by an ancient Greek divinity, but possibly also by a Moslem saint.

And one more source must be considered for his appearance: alchemy:

> But the image of the Green Man also pervades Western alchemical tradition as the figure of Mercurius, the Lumen Naturae, whose power for self-generation, self-transformation and self-destruction was described by the alchemists, who understood this energy to be divine life in all nature, ever changing, yet ever the same.
>
> (Baring and Cashford 1991 p. 412)

In the background of all this stands the figure of the Great Goddess, whose worship extended five times longer than any of the others since. She was the purveyor of all life; human, animal and vegetable. Her worship was not eradicated with any ease; the Church was reluctantly forced to acknowledge this, and so Mary was brought into focus, named Theotokos, Mother of God, and had superimposed on her much of the trappings of the worship of the Great Goddess.

In the Gothic cathedrals, as Antoninus had told me all those years ago, the churches named Notre Dame were mostly built on the old temples and shrines of the Goddess. And there we can find two indications of the ancient cults. One is the preponderance of statues of the Black Virgin, mostly brought back from the Holy Land. Writing of the Gothic cathedrals, Gary Lachman notes that "A common feature of many of the churches is that they possess statues of the Black Virgin, which are believed to have miraculous powers" (Lachman 2015 p. 187). In all the material written about these statues, scattered all over Europe, guesses have been made about why they are black – everything from ethnicity (Mary was Semitic) to the statues being blackened by the smoke of church incense over time has been suggested. My own guess, which seems to be solitary, is that the carvers of those particular statues, in their bursts of creativity (thought by the Greeks to be given by Dionysus), touched on a memory deep in their DNA, of the black mother eons ago in Southern Africa, the Mitochondrial Eve, Mother of us all. Always there is a magical connection to these Black Madonnas, thought of as providers of miracles, magic. This belief even traveled to the New World, where in Mexico the miraculous Virgin of Guadalupe is dark-skinned, though this is attributed native Indian ancestry. She appeared in a vision to a peasant, and when he could not convince the Bishop of what he had seen Guadalupe filled his cape with roses, which assured the Bishop that there had been a miracle. A large cathedral has been built at the site of the miracle, though there are those who still think that it was an appearance of Tonantzin, the Aztec Goddess whose temple was there.

In the museum in Mexico City stands a possibly even older Goddess, Coatlicue. Her carved size is enormous and I was overwhelmed, standing in front of Her. Particularly when I saw that where Her face should be there were actually two serpents' faces, facing one another. This was a surprise, and confusing, until I realized that it was the DNA symbol once more, as realized in ancient Greece as the caduceus. Snakes were always companions of the Great Goddess. What secret knowledge could date back millennia to present Her with that face? Knowledge that shows up in creation myths all over the world in diverse cultures in the use of the imagery of double snakes, two ribbons, ladders and vines. And these also appear in shamanism. The same symbol appeared for the healing gods of Greece, and still does into modern times. So, we are confronted with an archetype that is possibly the most powerful in our world. And She keeps appearing even when Church, and even State, forces try to eradicate Her. She was certainly the driving force, along with Her son, behind the building of the Gothic cathedrals.

An even more mysterious figure related to Goddess worship is to be found in many churches all over Europe: "the appearance from the early 12th century of the figure known as the Sheela-na-gig. Found frequently in Ireland from which her name comes...and surviving in scattered concentrations from England to Czechoslovakia" (Anderson 1988 p.128). This is the figure of a small, nude woman, sometimes smiling, sometimes looking menacing, her hands spreading open her vulva. To see her stone-carved figure in the midst of a Catholic church or cathedral is surprising. However, in terms of Goddess worship, she is a figure that dates from millennia back. I believe these carvings are more modern conceptions of Baubo, who in the stories of Demeter was the one who, by lifting her skirts in front of the Goddess and displaying her genitals, caused Her to laugh in the midst of grieving in Her search for Her lost daughter Persephone. There are even indications that Baubo made an appearance in the rituals enacted in Eleusis. I believe this figure can be traced back further, to matriarchal times when female sexuality was seen as something marvelous and awe inspiring:

> the miraculous nature of the vulva seems to have taken hold of the imagination of Paleolithic humanity…the vulva is the magical wound that bleeds and heals itself every month…woman bleeds but does not die and when she does not bleed for ten lunar months, she brings forth new life.
>
> (Thompson 1981 p. 109)

What is surprising, then, is that this figure found its way into the sancta sanctorum of the Roman Catholic Church, which has consistently cast into the darkness any reference to the lower parts of the human body. The masons and carvers, however, were from the very peasant classes that had held onto the old religion, disguising it to avoid open conflict with their masters.

The result of this intrusion of the matriarchal archetypes into the very heart of the Church was perhaps best summed up by Leonard Shlain:

> As literacy increased, the extreme emphasis on feminine values that had inspired the ethics, customs and mythology of the Dark Ages lingered on. Between 1000 and 1300, during the period known as the High Middle Ages, Europe enjoyed a Golden Age during which feminine values enhanced by orality teeter-tottered into relative equilibrium with masculine ones encouraged by literacy.
>
> (Shlain 1998 p. 293)

Simultaneously with this, and underlining it, "Toward the end of the eleventh century, Europe saw the mysterious appearance of the troubadours, Provencal poets who sang of their idealized love for an unobtainable woman" (Lachman 2015 p. 187). The author goes on to speak of the links between them and Sufi poetry, which speaks of love for the Divine erotically. Once again, Islam – the Islam that as we have seen before, held the Western birth records and had studied them.

The earliest appearance of what came to be called the troubadours was the nobleman called The Troubadour, William IX, Duke of Aquitaine and Count of Poitou. He was not necessarily the first troubadour as it has been agreed that the work of his that survives is a very refined example of the genre, suggesting it was already in place before him. William IX lived from 1071 to 1127.

The troubadours themselves were from Southern France and Northern Spain, and mostly wrote in Occitan, or *langue d'oc*. This area was a country unto itself made up of small fiefs stretching from Aquitaine through Carcassonne, Auvergne, Montpelier and Toulouse, among others. And it had the peculiarity at that time of having some of its fiefs in the hands of women. In Occitania, women – aristocratic women, it must be said – were granted a privileged status not common to their sisters throughout the rest of Europe. This was due, at least in part, to Theodosius Code of 394–5 which gave sons and unmarried daughters an equal share in their father's estate, and the Code of Justinian, which gave the rights of his wife's dowry to the husband to use, but not claim. Both of these succeeded in providing the Occitan woman of aristocratic birth a great deal more freedom than was afforded other women throughout Europe. One outstanding example of this was Ermengarde of Narbonne, who ruled over the land of her father when he died, and rode into battle too. She was also Patroness to some troubadours and had correspondence with one, a female by the name of Azalais de Porcairagues.

It was from the ranks of the returning Crusaders, particularly those of the doomed First Crusade at the end of the eleventh century, that the earliest troubadours came. They came, as has been previously written, bearing not only the souvenirs of the Holy Land, but also ideas and customs experienced there. In addition to his own work, within the journals of William IX, who went forth into the Second Crusade, were found fragments of a few Arabo-Hispanic verses copied by his own hand. Possibly these came from his forays in conjunction with his allies in Castile and Lyon in order to push back and crush the Moors of Cordoba. Amongst the prisoners, some of whom he brought back to Southern France, it is obvious he might have come in contact with the then developing Sufi mystical poetry. These Sufi Islamic musical forms always contained reference to The Beloved. Ezra Pound in his Canto VIII proclaims that William "brought the song out of Spain/with the singers and veils" (Pound 1996). Only eleven of William's own poems remain.

The theme of the troubadour poetry was courtly love for the unattainable woman of high rank, usually married, and thus beyond his reach. In some of the poetry there are some sexual innuendoes. All this was anathema to the Church, of course, with its continuing hatred and fear of women, and which was pressing for devotion to the Virgin Mary, who they proclaimed was Herself of virgin birth. The songs of the troubadours elevated woman herself to a position superior to man.

What we hear less of are the twenty or so women, writing under their own names, who were the trobairitz: the female troubadours. More than twenty have been identified, and much of their work has been translated. "The women's [trobairitz's] motivation for writing...is more often urgently serious than is common among the male writers; it is a response springing from inner needs, more than artistic, didactic inclination" (Dronke 1985 pp. 7–8).

It sounds to me that the voice of the trobairitz was Dionysian, a feeling cry for balance in love as well as life.

It was perhaps all this: the poetry, the elevation of women (at least aristocratic ones), the intrusion of Middle Eastern ideas, coupled with the heresy of the Cathars. This led to the Albigensian Crusade, which called upon European knights to obliterate the Cathars. Among the beliefs and traditions of that faith that ran contrary to those of the Church, it allowed women equal status as Perfects, priests of that faith. By 1244 it was done, as were the troubadours, many of whom had Cathars as patrons and connections.

In 1231, the Pope, perhaps sensing and fearing the invasion of the feminine and the Dionysian, appoints the first Inquisitors "of heretical depravity," and gradually women and the feminine are once more put in their place. As were the Arabs and Jews in Spain, along with all their vast knowledge, later driven out of Europe.

All the anti-woman rhetoric was at hand in the *Malleus Maleficarum*; "All witchcraft comes from carnal lust, which is in women insatiable... wherefore for the sake of fulfilling their lust, they consort even with devils" (Shlain 1998 p. 366). And if it is thought that 300 years later, with the Reformation and the coming of Luther and Calvin, the attitude as regards women had changed, think again. With Calvinism "Women were harshly regimented. They were forbidden to wear lace, rouge, jewelry, or fine, colorful clothes" (Shlain 1998 p. 334). And "Fornication (sexual relations before marriage) was punishable by exile or drowning. Adultery – death; blasphemy – death; idolatry – death; heresy – death..." (Shlain 1998 p. 334).

References

Anderson, William (1988) *The Rise of the Gothic*. New York: Dorset Press.

Baring, Anne and Cashford, Jules (1991) *The Myth of the Goddess: Evolution of an Image*. London: Penguin..

Cheetham, Tom (2005) *Green Man, Earth Angel*. Albany: State University of New York Press.

Dronke, Peter (1985) *Women Writers of the Middle Ages*. Cambridge: Cambridge University Press.

Lachman, Gary (2015) *The Secret Teachers of the Western World*. New York: Tarcher/Penguin.

Pound, Ezra (1966) *The Cantos of Ezra Pound*. New York: New Directions.

Shlain, Leonard (1998) *The Alphabet Versus the Goddess*. New York: Penguin/Compass.

Thompson, William Irwin (1981) *The Time Falling Bodies Take to Light*. New York: St. Martins.

Toynbee, Arnold (1978) *Choose Life*. Oxford: Oxford University Press.

6   Women mystics

In the eleventh century, what must be recognized as an early "women's movement" must be looked at as being just that – a move by women to express a thinking that was a woman's own. Though it developed along Christian lines, Christianity being the only accepted religious expression in the West, it was not of the Church. Somewhere around 1210, a group of women appeared first in Liege (now in modern Belgium) and spread rapidly to France and Germany. They called themselves Beguines, and though they never precisely organized, never created convents and never adopted a rule, the idea itself spread. They did not take permanent vows, though their stated intent was to live lives of chastity and poverty in the manner of Christ. Some of them lived in groups with like-minded women, some on their own and some even in a family. They went out in the world to make their livings, generally by embroidering Church vestments or weaving, and always working with the sick and poor, with some even founding hospitals.

They lived in direct contradiction to the norm of the times, which was for women to live under the rule of a man as a wife and mother, or to go into a convent as a nun. The Church was highly ambivalent toward them, as could be expected, and of course they never received permission from the Pope to live in the manner they chose – freedom. At times they were accused of heresy, and a few were even burned at the stake. But there were thousands of them known to the poor for their good works, and the lives they lived were within the approved lines of the Christian faith. Thus, they were granted special privileges and given exemptions denied the orthodox religious orders.

They celebrated their religion in song and in dance, performed the male-only accepted tasks of translating Holy Scripture, and even had debates over theological issues. Many were mystics, experiencing ecstasy, going into trance and having visions ... this has echoes of the maenads.

The Beguines gradually died out; the last reports of such communities were in 1969 in Belgium and Holland.

Among the three mystical women discussed here, only one was a Beguine. Another was a nun who became a saint in the Church, and the third, an anchoress. But all experienced an inner relationship to the Divine and left written works to share with the world to come. Alongside all the anti-woman rhetoric of the twelfth and thirteenth centuries, there arose the voice of some women within the Roman Church, some in various religious orders that were arising, that can only be described as "mystical," as well as a few not strictly within the Church. By this I mean religious women who were going down deep within themselves in meditation and prayer and experiencing the Divine within. This was a reflowering, I believe, of the experience of the mystery religious experience as in the Dionysian and Eleusinian rites of the ancient world, where the Divine was not just found someplace "out there," but was also to be found within each one of us. The Church was always wary of these voices, both male and female, fearing perhaps that the paternal rulings of the Roman hierarchy might be in jeopardy given the individual experiences expressed.

It could be said that the shift to the honoring of the inward experience was in some measure due to the preaching of St. Bernard of Clairvaux sometime in the twelfth century. His sermons and preachings spoke of the mystical union with the Divine in terms of lover and beloved, much as the Sufis, the mystical sect of Islam, did. Interestingly, this was also the time of the emergence of one of the great Sufi teachers and philosophers, Ibn Arabi, native to nearby Spain, also in the twelfth century, who preached a similar relation to God. The relationship of the Roman Catholic Church with God prior to this time was more of a patriarchal, almost militaristic love and comradeship. By bringing love into the equation, the stage was set for a more feminine relationship.

Across Europe, the writings and activities of women in the Church were getting attention. Their writings have relevance still for us today. And in those centuries those writings attracted and opened more choices within religious orders for women. No, women were still not performing priestly duties, but because of this they had more time for meditative work and prayers, allowing them the deeper experience in which they found the core of the Divine within themselves[1] – and even speaking of visions of "God the Mother," although still using the masculine pronoun "he."

A few examples of women mystics of the time are currently being brought to light by certain voices within the Women's Movement here in the twenty-first century.

Mechtild of Magdeburg

Mechtild was a Beguine who was born into a knight class family in Saxony. At only twelve years of age she had her first mystical experience, which she later described as a greeting by the Holy Ghost. These continued, virtually every day. At age twenty, she moved from the comforts of her class to Magdeburg, where she lived with a group of women who identified themselves as Beguines.

Her confessor, a Dominican priest named Heinrich of Halle, encouraged her to write down her vision as instructed by God when she was forty years old. She dictated these to her confessor in Low German, for although educated she had no Latin and, indeed, at the time it would have been unusual for a woman to write. Heinrich was a great supporter of her, though she had many enemies given she was a woman not within a Roman Catholic order. The work, often exquisite imagery, was entitled *The Flowing Light of the Godhead*, and often housed in erotic terms.

For the next twenty years, she continued to receive these visions and revelations and dictate them to Heinrich, who organized them into six books, all in her Low German. It was the first book of the time to be written in the vernacular rather than Latin.

However, many within the Church questioned the truth of her visions and conversations with God, and her book was, at times, threatened with burning. When she spoke to God about this, the answer came that that the truth cannot be burned. Through these challenges she developed a jaundiced view of many of those within the Church hierarchy, calling them corrupt and sinful themselves and, thus, no judges.

In her visions, she saw herself not only as the Bride of Christ, but also as his partner in suffering through the many illnesses she endured, as well as the constant censure of her foes within the Church. Early on, she self-imposed fasting and even flagellation, though later it was her illnesses that martyred her.

At about 63 years old she entered the Benedictine Convent at Helfta, not as a nun but as the honored guest of their enlightened Abbess Gertrude. She was seriously ill by then and going blind. It was there that she dictated the seventh and final book of *The Flowing Light of the Godhead*. Her date of death is disputed, with some scholars citing 1282, others 1285 or even 1297.

Julian of Norwich

A contemporary of Chaucer, Julian of Norwich (1342–1416) was the first woman that we know of to write a book in English. Her introspections of the Divine and its life in us and our world have become known as *The Book of Showings of the Anchoress Julian of Norwich*, also known as the *Revelations of Divine Love*. A simple title, perhaps, for some profound insights and what we might otherwise call "visions" as opposed to "showings." This was the essence of Julian – simplicity and modesty. An anchorite, or anchoress as the Church styles it, she lived a great deal of her life in seclusion in her cell attached to the St. Julian's Church in Norwich, England.

Very little is known of her background or early life in Norwich, a town ravaged by the Black Death. She herself was quite ill, and thought she was dying when she began receiving visions in her thirties. Her "showings," as she referred to them, were of the Passion of Christ, and of Mary, the Mother of Jesus. And in the course of writing down her visions she refers to God as "the true Mother of life and all things. To the property of motherhood belong nature, love, wisdom and knowledge, and this is God" (Colledge and Walsh 1978 p. 60) – a vision of God as feminine which was unusual, to say the least. Such a metaphor identifying the very masculine Jehovah with the feminine Mother is still unique.

In addition to the writings she made of her visions, Julian was also important to her community. She obviously did not live the completely secluded life one associates with anchorites, for in addition to her prayers and good works for the poor, she was regarded as an authority on spiritual matters and often served as an advisor. The very well-known Margery Kempe, another mystic of the time, visited Julian and spoke with her. This is written of in Kempe's *The Book of Margery Kempe*, thought to be the first autobiography written in English. Another first, also by a woman. She writes of the good advice she received from Julian.

Julian was most probably not a nun, and is thought by some to possibly have been a young mother who lost her family in the plague that swept Norwich and much of England. She has never been given sainthood by the Church, though the Anglican Church has accorded her a feast day on May 8th. The Roman Catholic Church, as of 1997, has put her on the waiting list to be declared a Doctor of the Church; she is recognized as a holy woman of God, and often referred to as "Mother Julian" or "Blessed Julian."

However, there are other women mystics who have been canonized.

Saint Teresa of Avila

One such was Teresa of Avila (1515–1582), who entered a Carmelite Convent, very much against her father's wishes, at twenty years old. Her father, a Jew at a time when Jews were demonized in Spain, became a converso or forced convert, as he decided to live there too. He had two children by his first wife, who died after their birth; he then married a young woman and had many children by her, Teresa being the first. Among the rest were seven sons, most of whom became conquistadores in the West Indies.

Teresa was said to be a lively, vivacious girl, outspoken and sociable. She was described by contemporaries as "lovely," with dark hair and eyes. She herself wrote later of this time of taking great care with her dress, ensuring she was scrupulously clean, and wearing the finest perfumes. She admits to being a people pleaser and enjoying attention. Living very close to her cousins, there were even rumors of a romance with one. Evidently this reached the ears of her father, who sent her to a strict convent school, Our Lady of Grace.

Once there, she found that she enjoyed the peace it brought her and began to wonder if she should become a nun. She prayed to God to send her some sign that this was her vocation, but none ever came. She became ill at school, suffering from high fevers and collapses.

Her father brought her home at this point and she acted as a surrogate mother to her younger brothers and sisters, who were now motherless. Several years of this ended when she became convinced that she did have a vocation as a nun. Possibly the example of her mother's constant childbearing, which had driven her further and further into being a recluse, and then an early death, set the stage for Teresa's choice of the exact opposite: a nun in a sacred, closed order – the Carmelites. She then entered the Convent of the Incarnation, though this was strongly opposed by her father, who only relented a few years later when it became obvious to him it was her vocation.

She was perhaps best known for her reform of the Carmelite order, feeling poverty should be a rule for both male and female orders. However, she made sure that each nun in her convent had her own cell, regardless of family wealth or background. And flagellations, such as mortification, were not permitted. She was protective of the women in her charge. Her zealousness and writings drew the attention of the Inquisition, which focused particularly on women and Jews, or suspected Conversos who secretly practiced their Jewish faith. They never were able to find any indication of heresy in Teresa, her writings or her practice.

In her thirties, she again experienced a debilitating illness, and once again came home. Even so ill, her prayer life flourished, and she speaks of this in her autobiography. She was undergoing a strenuous treatment by a famous healer known to her family, but finally fell into a coma.

Though she was believed to be dead by all around her, her father would not allow anyone to touch her. After four days, extremely weak, but alive though unable to walk, she insisted on being brought back to the convent where she preferred to be to die or recover. Slowly, after a period of two and a half years, she was once again able to walk and stand. A miracle, said some. It was at this point that she experienced what some have said was a second conversion, at forty years old. A new statue of the crucifixion had been placed in the convent, and as she passed it one day she found herself down on her knees, crying hard, saying she was guilty of twenty years of ignoring His pain and begged forgiveness.

Her writings of her visions and her thoughts were particularly powerful. Her Interior Castle and The Way of Perfection are read today and cherished.

Before she died at 67 years old she had founded as many as seventeen convents and four monasteries in her vision of the order as being based in poverty and love of God.

Note

1 A paper, recently presented at the Wijngaards Institute for Catholic Research, claimed to show three ancient artifacts from the fifth and sixth centuries portraying women in leadership roles at three important Christian churches: St. Peters in Rome, the Church of the Holy Sepulchre in Jerusalem and the Hagia Sophia in Constantinople.

Reference

Colledge, Edmund and Walsh, James (1978) *Following the Saints: May 1st through August 31st*. North Carolina: Goodwill Publishers.

7   Dionysus underground

The horrors of the Inquisition, widespread by the Church, drove Dionysus even further underground, where He appeared in several different guises throughout the next centuries. Not just to the "common" people, Goddess worshippers in secret, but also to the intellectual cognoscenti, who had great misgivings about the teachings of the Roman church. Often they disguised this by studying one thing and calling it another so as to fool the Church. Most often they mentioned it not at all. As early as the fifteenth century, a well-known benefactor, Cosimo de Medici, sensing a slight scholarly relaxing of the Church, asked the head of the Florentine Academy, Marsilio Ficino, to set about translating a set of as many as seventeen manuscripts that had been found in the Middle East.

The publishing of these in 1471, entitled *The Corpus Hermeticum*, thrilled and excited scholars and artists all over the world, who regarded them as ancient wisdom from Egypt, ancient Greece and even Jewish traditions. The "hermeticum" comes from the name of one of the greatest, most highly regarded, possibly mythical magicians of all time, Hermes Trismegistus (or Thrice Greatest), now regarded as most likely coming from the first or second centuries of the Common Era. One of the perceived heresies of the hermetic path was that it stated that all humans contain a divine spark of power. With this belief, then the authority of the Church, and its priests, is not required; it is within ourselves to work toward. As in the Dionysian Mysteries, "the spiritual spark which is a residue of the god, remains in us" (Lachman 2015 p. 73). Dionysus was the central figure in His own mystical rites and one of three in the Eleusinian Mysteries. Both these mystery religions provided a personal encounter with deity, though ritual and a priestly caste were both part of it. However, the Initiate of both of the cults was most impressed with their own encounter and perception of divinity. And, in reading some of the short, intense writings of famous initiates, one gets the impression that the visions were life changing.

To avoid Church censure and punishments, as stated above, various names were given to these studies. Mostly the subjects were said to be merely translations from earlier traditions and pains were taken to make links, some tenuous indeed, to accepted Church beliefs. In this way, scholars all over the world were learning that there were a multitude of ways to approach the Divine, on one's own, through study and thought. The translation of *The Corpus Hermeticum* was just the beginning. And Dionysus was present: "Of all the gods of ancient Greece, Dionysus is the one who still remains within us…" (Lachman 2015 p. 73).

I believe this is because we are all looking for more life, more feeling of being truly alive, and too often it is the rules and regulations side of organized religion that takes this from us. The earliest, Goddess-based religions were all about life and sexuality. Too often the focus now is on practicing denial, particularly of our natural instincts in our lives, in promise of some glorious afterlife. And Dionysian sexuality is particularly feared, I believe, even today. An essential and characteristic feature of the world of Dionysus, which is extensively documented by iconography, is precisely sexuality: not an expression of fertility or reproductive inclination, but of joie de vivre – that is to say, a sexuality which is "useless" and "playful" (Isler-Kerényi 2007 p. 23)

The "Rosicrucian furor" began in 1614 when a strange document announcing the Brotherhood of the Rosy Cross appeared in Kassel, Germany. The Brotherhood proclaimed the coming religious, scientific, social and political reformation of Europe, couched in strange Hermetic and Alchemical language. In an immediate sense it meant liberation from Catholic and Hapsburg rule.

(Lachman 2012 p. 22)

And what of the women in this time? Aside from those who the Inquisition named witches and burned at the stake, the wise women, in whose hands births were aided and herbal medicines mixed and used to heal, were deemed unfit for such practice, and what has become known as the medical profession was gradually given over completely to a masculine establishment which remained solely in the hands of men until the last 100 years. Only the midwives in some countries were allowed to keep practicing under male guidance.

A parallel can be drawn between the sexual repression of the maenads at the end of paganism, the repression of witches by the sexually demented Inquisitors of the Middle Ages, and the repression of female sexuality in the nineteenth century by Puritanism, which classified as hysteria every sort of feminine rebellion of a sexual nature.

(Paris 1990 p. 33)

In between the Middle Ages and the nineteenth century, some mention must be made of Franz Anton Mesmer. Mesmer started out as a graduate in medicine in the late eighteenth century, and he was the progenitor of what came to be known as animal magnetism. He was disenchanted with the medical practices of his day and searched for a more therapeutic procedure. In his studies of ancient medicine, he came across the idea of there being effects on humans from the astral realms and developed a treatment to align his patients with a universal fluid. This was the beginning of what has come to be called mesmerism. He was so successful with this in Vienna, where he practiced, that the hostility from the other doctors resulted in his moving to Paris to continue in what he felt was a more open atmosphere.

Once in Paris, instead of continuing work with individual patients, he employed a group method which became, for some reason, more and more theatrical. He began to wear robes and even suits of a lavender color, and carried a magnetized rod or wand. He attracted many more women than men, and he mesmerized them into a sleep from which they claimed to awake refreshed and healed. It wasn't long before the French authorities began to pay attention. There were questions about whether this "refreshed" state was actually some sort of sexual gratification – and once more the connection between women and sex was seen as a threat, particularly to the conventional medical establishment. Had the maenads returned? Within a few years, his medical colleagues found a way of shutting him down. Mesmer faded into obscurity, but his findings in mesmerism were continued in more scientific and discrete ways by others. And hypnotism became a favorite on stage and at parties, too, right into the nineteenth century.

While the Western world was continuing the debate over Mesmer and his method, in 1848, in a very unlikely setting, what would become a major religion was having its earliest beginnings. Another form of trance was the basis. Two young sisters in the farming community of Hydesville, New York, confided in a neighbor that there were strange things happening in their small, shared bedroom in the farm house. Each night, they disclosed, raps and bumps were heard coming from the furniture and the walls of their room. They felt these might have to do with an otherworldly intelligence or ghost. Skeptical though she was, the neighbor agreed to come over that night, and, along with the girls' parents, went to their room at bedtime. The little girls huddled together on the bed while their mother gave a demonstration of what happened each night. At her command to count to five, the walls of the room trembled with five heavy thuds. Then she ordered fifteen, and that number of raps followed. Next, she asked the "spirit" the age of their neighbor, and it responded with thirty-three raps – her correct age. Finally, the mother asked that, if this was coming from an injured spirit, to reply with three raps, which it did.

Frightened by the seeming discovery of a disembodied spirit, the parents sent the two little girls to live with their older sister in Rochester, New York – a move that was to bring the Fox sisters into unexpected prominence.

Rochester at that time was seething with reform and religious activity. Mormonism was born there. Some prominent members of the community heard about the girls and the phenomenon they experienced at home. Somehow a story attached itself that in that same farm house, years before, a peddler had been murdered. A delegation was sent to the farm house and, in the basement, bone fragments and human hair were found. That was all that was needed for them to want to test the girls further. What was now called a séance was held in Rochester at the home of one of the prominent citizens, and those in attendance, even the skeptics among them, were impressed by distinct raps from the floor in response to questions asked. The girls, to prove the reality of what even the skeptics among the crowd had heard, were then examined in private, and even disrobed. No evidence of a hoax was found.

An American man who claimed to be a seer heard of the Fox sisters and brought them to where he lived in New York City to test them for himself. And thus was Spiritualism born. It attracted many who were increasingly dissatisfied with the Christian religions which denied them having much of a hand in their own spiritual lives. Again, some of the concepts in *The Corpus Hermeticum* were given new light. And of course, Spiritualism offered them the chance of speaking with those loved ones who had crossed over. In the midst of this, a great many decried it all as a fraud. In fact, the Fox sisters, at various times, said it was a hoax themselves. As late as 1888, the sisters publicly decried the whole thing as having been made up by themselves as children.

But before then had been the American Civil War in which so many lives were lost, and many of their families and loved ones had attended séances where they were sure they had spoken with their deceased loved ones. The sister's "confessions" were not taken to heart by those who still believed. And Spiritualism survived.

In the midst of all this Spiritualism, another figure appeared, a woman around whom controversy still swarms: Madame Blavatsky. She was one of the most controversial women of the late nineteenth century and I feel is best introduced by the following excerpt:

> ...and it is generally agreed that her Theosophical Society...was more or less the official starting point of the modern spiritual revival. By "modern spiritual revival", I mean our contemporary widespread interest in a direct, immediate knowledge and experience of spirituality.
>
> (Lachman 2012 p. ix)[1]

Helena Petrovna von Hahn was born on August 12, 1831, in Ukraine. Both she and her mother lived through the birth, though both were sick with cholera and barely survived. Her mother was a respected literary writer from a noble family, married to a military officer. As many military families still do, they lived in many places because of her father's career. However, at one point her mother decided enough was enough for a time. She took Helena and her siblings to live with her own father, who was a trustee for the Kalmuck Buddhist Tribes of Astrakhan, on the Volga River where it ends in the Caspian Sea. It was here, I believe, that Helena discovered and assimilated the beginnings of her lifelong dedication to Eastern mystical thought. She was specifically exposed to the Mongolian Lamaic system, which provided her first taste of Tibetan Buddhism. Whatever she learned and absorbed during her year-long stay there was later reinforced when living once again with her father and mother in another location, where she attended a summer camp of Kalmuck Buddhists, some of whom she had met before. There, she learned the properties of the Tibetan Prayer wheel, and it was at that time she began to speak of a "protector" who appeared in her dreams – perhaps one of the Masters of whom she later spoke and wrote.

In her early teens, she lived in proximity to the library of her great-grandfather, Prince Pavel Dolgorukov, and began spending time reading there, which she later said contained hundreds of books on alchemy, magic and other occult subjects. She claimed in a letter written to a friend, years later, that she had read all of them by the age of fifteen. A couple of years later, still young, after an almost immediately failed marriage to a man several years her senior, she ran away to Europe, emerging as Madame Helena Blavatsky.

From Europe, she traveled all over the world, including, she would later claim, several years in Tibet. It was this claim that caused so much controversy as in the nineteenth century it was thought that no white person had ever been admitted to that country, let alone a woman. But she was adamant, and never retracted her claim.

Once back in Europe, she read of the Spiritualism craze that was conquering America and decided to go there. She went straight to the author of the articles she had read, Colonel Olcott, and the two immediately became close friends and colleagues. Though not accusing any of the spiritualists she met as being frauds, she claimed that they were not contacting the spirits of the dead, but, rather, other denizens of the astral realms. She began to speak of her own meetings and contact with those whom she called the Masters. These were not disincarnate spirits, but men of high spiritual standing all over the world. And she brought popular knowledge of Buddhism to the West. Yes, she could apport – that is, bring objects into being which had not been present before – and she did this many times in front of both believers and non-believers. But as time went on, she did it less and less as she felt such displays were "beside the point" – that is, they did not contribute to the spiritual growth of those who did it, nor to that of the viewers who seemed bent on proving there was some trick to it all. (It is interesting to speculate that a few of these wound up at Stanford University, where a room full of such purported objects was still around in the 1970s, being from the collection of the brother of the founder, Leland Stanford.)

Her commanding presence, her enormous obesity, which she sheathed and wrapped in exotic Oriental fabrics, and her sometimes downright rudeness both attracted and repelled those with whom she came into contact. One wonders whether, had she been male, those very characteristics would not have been seen as somewhat admirable. The fact remains that the two monumental works she wrote, *Isis Unveiled* and *The Secret Doctrine*, had a great impact on those who could get through the enormous task of reading them, as jam-packed with occult information and ideas as they are. And the Theosophical Society, which she and Olcott founded in 1875, still remains a powerful voice for the Brotherhood of Man, and for directly received and experienced mystical experience and its universal application – once again, the mystery religions of old brought forth into modern times by a woman.

This has been the thread running through history along with the repression of the Dionysian. One of its strongest manifestations in the late nineteenth and early twentieth centuries came from Sigmund Freud, pioneer of the early days of psychology.

Freud became interested specifically in hysteria through the work of Charcot and of Breuer. A report that he presented before the Viennese Medical Society was met with negative reactions, particularly as he maintained he saw hysteria in men also. One of the doctors present voiced concern over this, stating out loud, "But my dear Sir, how can you talk such nonsense? Hysteron...means the uterus, so how can a man be hysterical?" (Veith 1965 p. 263) – once again, the threat of female sexuality showing through.

The association of hysteria and other particular diseases attributed to women began as early as Ancient Egypt in patriarchal medicine. The earliest document discovered so far comes from about 1000 BC. It is only a fragment, named the Kahun Papyrus after the town in which it was found: "They were evidently part of a small treatise describing a series of morbid states, all attributed to displacement of the uterus" (Veith 1965 p. 2).

And, thereafter, hysteria has been identified as a female disease.

Freud came to believe that sexual repression, among other sexual problems, was at the basis of all neurosis, and even wrote "I do not think I am exaggerating when I assert that the great majority of severe neurosis in women have their origin in the marriage bed" (Veith 1965 p.267). Again, however, hysteria presented as a woman's disease attached to her sexuality – and continues to do so.

A Yale study was recently done on adults about their reaction to the pain of boys as compared to that of girls:

> The findings add to existing research showing that female pain is dismissed and undertreated compared to male conditions. This sort of dismissiveness of women's pain reflects a stereotypical belief that women are hysterical, and therefore if a woman is expressing pain, she must be exaggerating.
>
> (Goldkill 2019)

The author goes on to say "The results could show that women, who often face a lifetime of prejudice in healthcare, starting experiencing such prejudices from birth."

It seems obvious to me that this is a continuation of the fear of women based originally on their sexuality, as personified in the earliest manifestations of Goddess worship – seeing the feminine as divine. It appeared recently in the resolution of a court case here in the United States, where a 56-year-old man was being tried for sexual abuse of two teenage girls. The judge ultimately decreed that the 13- and 14-year-old girls were responsible for enticing the man to have sex with them – the mature man obviously unable to think through what he was doing with minor girls when presented with their sexual natures. The repression of the Dionysian is again linked with the fear and hatred of women.

Note

1 Gary Lachman's biography of Blavatsky remains, for me, the most balanced of all that is known about her to this day, presenting both the negative and positive takes on her and her claims, and obviously deeply researched.

References

Goldkill, Olivia (January 29, 2019) *Yale Study Shows Adults Take Girls' Pain Less Seriously Than Boys*. Quartz.com: https://qz.com/1535889/yale-study-shows-that-adults-take-girls-pain-less-seriously-than-boys/.

Isler-Kerényi, Cornelia (2007) *Dionysus in Archaic Greece: An Understanding through Image*. Boston, USA and Leiden, Amsterdam: Brill Academic Publishers.

Lachman, Gary (2012) *Madame Blavatsky: The Mother of Modern Spiritualism*. New York: Random House.

Lachman, Gary (2015) *The Secret Teachers of the Western World*. New York: Random House.

Paris, Ginette (1990) *Pagan Grace*. Dallas Texas: Spring Publications.

Veith, Ilza (1965) *Hysteria: The History of a Disease*. Chicago, Ill.: University of Chicago Press.

8 Dionysus in projection

In societies that hate and fear women, this becomes most apparent in how they seek to control them, particularly their bodies. The ancient matriarchal religions were rooted in the body of woman – it was the source of her power. As mentioned earlier in this work, for Paleolithic people the female body was a source of wonder and mystery. On the island of Malta, temples to the Goddess in the shape of her body have been excavated by archaeologists. In the rituals of the Great Goddess, an essential part was in the mysteries of the blood.

As patriarchy took over, as we saw in Greece, the female body was increasingly given a masculine slant, particularly in the case of the Goddess Athena, whose patriarchal myth included being born from the head of Her father Zeus and fully clad as a warrior. As was said in a lecture I attended on myth, "In all things she was for the male, except in marriage" (Ponce 1980). Apollo's sister, Artemis, was well equipped as an archer. And when the Goddesses were not thus masculinized in some way, they were often depicted with hideous aspects, like the Medusa, or greatly feared for "witchy" powers, as Hecate.

Many of the objects and aspects of the Great Goddess became associated with Dionysus: "Virtually all the Dionysian characteristics mentioned: figs, bulls, Muses, the Moon, dance, music, moisture, sexuality, regeneration of the earth, cultivation of plants…were originally under the aegis of the Goddess" (Shlain 1998 p. 139).

The one Goddess in Greek religion who remained virtually inviolable was Demeter – and, through Her, Persephone, Her daughter and Queen of the Underworld. Besides the rites of His own cult, this powerful mystery religion, which became a threat to Christianity, was the only other place in Greek culture where Dionysus could be found. Of course. He was the Son of the Mother (Harrison 1992 p. 402).

As women became more and more marginalized, so too did Dionysus as God of Women, who as time went on became known mostly as Bacchus, God of Wine. By some modern scholars, this is termed the Apollonian/ Dionysian split; by others, the Left Brain/Right Brain dichotomy, which has been greatly felt by women for more than 2000 years, but perhaps more acutely since the seventeenth century and industrialization.

In this split, in addition to marginalizing women, another aspect, just as destructive and also based on fear and hate, has been the projection of the qualities of the Dionysian, viewed as inferior and not quite "nice" to the predominately Apollonian Western European culture, onto other cultures, thus marginalizing them. This was intensified by the colonization all over the continents of Africa, Australia and North and South America. And, as said in a lecture I attended, "We must be about the business of withdrawing projections. Projections have, in essence, created much of what we live within" (Ponce 1980) – and, I would add, much of what other cultures projected upon them by Western culture, have had to live within.

A great deal has been written in the last seventy-five years about the gross indignities and barbarism inflicted on the many countries colonized and conquered by Western nations. The Europeans were so convinced of their superiority, and that the people of the lands they colonized had nothing of value to teach them, that they destroyed their cultures.

> More than 80 million people are estimated to have been living on the southern and northern continents of the Americas in 1492; within three hundred years, the "explorers", "conquistadores", "colonists", and "pioneers" – exterminated the majority of the native populations: their current number is approximately 10 million.
>
> (Shlain 1998 pp. 349–350)

This was the pattern of colonized countries all over the world. Their peoples – considered inferior and all the darkness of the conquerors projected upon them –committed suicide in large numbers as they saw the women and the landscape raped by the intruders. These same intruders who continue their rape in more subtle ways even today. Women raped of their power, and landscapes decimated in the name of progress.

All this is being brought to light more and more in this century, as we confront our misdeeds and, hopefully, our projections on the cultures we almost destroyed.

It was heartening to see, in one of its recent issues, the prestigious *National Geographic* magazine own up to its part in this. As one of the people hired to look over issues of race in its publishing history was quoted:

> until the 1970s National Geographic all but ignored people of color who lived in the United States, rarely acknowledging them beyond laborers or domestic workers. Meanwhile it pictured "natives" elsewhere as exotics, famously and frequently unclothed, happy hunters, noble savages – every type of cliché..
>
> (Goldberg 2018 p. 1)

My attention was drawn particularly to the African American culture which developed from the time they first came as slaves to the United States, and to the Gypsy or Roma culture that spread all over the world.

Projections went on long after slavery and colonization was over. Witness the fact that an estimated 500,000 gypsies were slaughtered by the Nazis in World War II, reducing their population by 70–80 percent in Europe by 1945. Despite these terrible prejudices, great art came out of it, though the pain comes through. "Andalusia, a region steeped in its Roman, Jewish, Moorish and Gypsy complexes" (Pedraza 1990 p. 58): it is interesting that three of the most reviled and projected upon cultures by Western society have come together to produce one of the most beautiful art forms within music, dance and drama – flamenco. Early on, one of Spain's greatest poets and playwrights, Gabriel Garcia Lorca, sensed that in its singular creative art, flamenco bore the imprint of what he identified as having "leaped straight from the Greek mysteries to the dancers of Cadiz or the beheaded, Dionysian scream" – the duende (Lorca 1955/1998 p. 60).

The first taste of flamenco that I had was in San Francisco's North Beach district in 1958, where, in a café serving Spanish food and drink, there was a troupe of young musicians, dancers and singers. Their performances reached something so deep inside me that I was unable to articulate it. I kept it buried inside me until years later when I encountered the poetry and plays of Lorca. Most important to me was his essay, "The Play and Theory of Duende." I realized then that it was a different sort of creative imagination that the finest flamenco artists sought. An inspiration not of the head, but one that rose up from the darkest parts of the psychological interior, from the Dionysian body. I believe that the dance movements in flamenco have evolved from the ecstatic movements of the maenads, Dionysus's bands of women followers who worshipped Him in frenzy.

In the poems of Lorca, especially his "Lament for Ignacio Sanchez Mejias," death is always present, as it was always with Dionysus – present, inevitable and a component of life itself. Death resonates in the repetitive rhythm and words of the one line that echoes in an incantatory and initiating way – "At five in the afternoon" – the hour of the goring in the bullring of the torero, the crying out of duende. When duende is present in a performance, the audience is equally involved. This poem is as powerful in English translation as in the original Spanish.

> At five in the afternoon
> It was exactly five in the afternoon.
> A boy brought a white sheet
> At five in the afternoon.
> A trail of lime ready prepared.
> At five in the afternoon.
> The rest was death and death alone.
> At five in the afternoon.
>
> (Lorca 1955 1998 p. 67)[1]

The next two stanzas have the same alternate line repeat of "five in the afternoon." The impact of those lines and the death of the matador is felt not only by the poet, but the presence of the duende and Dionysus is felt by the reader. Dionysus as Hades, Lord of the Underworld who observes and waits.

In the singing of flamenco, if the singer has invited duende into her/his performance, it is there also. One widely told anecdote among aficionados is that of a singer in a café in Andalusia who notices a famed critic of the art has come in and ordered a drink. So she sings out her best, most polished performance. At the end, the critic applauds slowly, without enthusiasm. So she sings another, this time forcefully directing it at him, and gets the same result. Finally, she is so distraught that she sings out from the depths of her very being, her voice rasping and at the end the critic stands and applauds loudly, crying out "ole!" The duende has made its appearance. So too in the story of a flamenco dancing contest in which an old woman stands up, striking the pose of the beginning of the dance, stamps her foot, and wins the contest on just that one move – once again the duende and Dionysus, this time appearing in old age.

"Lorca perceived a deep connection between the Gypsies, the oppressed Spanish women and the Blacks of New York's Harlem":

> You Harlem! You Harlem! You Harlem!
> No anguish can equal your thwarted vermilions
> your blood-shaken, darkened eclipses,

your garnet ferocity, deaf-mute in the shadows,
your hobbled, great king in the caretaker's suit.

(Cobb 1992 p. 111)

The connection that Lorca saw between flamenco and Black Americans shows in attitudes toward the jazz and blues that developed with that population. The criticisms and fears of the Western culture in the early days sound very similar: "Like early flamenco, early jazz had been associated with despised ethnic groups, gangsters, free-spending blue-bloods and hedonism" (Mitchell 1994 p. 46). And,

> ...the spread of American jazz is best understood by studying the reactions of those who were hostile toward it. Alarmed musicians, clergymen, journalists and even philosophers heard the new music as sensuous and indecent...its primitive rhythms aroused animal passions; jazz embodied a spirit of lawlessness and general revolt against authority.
>
> (Mitchell 1994 p. 45)

The fear of much of the American public directed at the Black population stems from the early days of the slave trade when thousands of Africans were transported in the most horrendous conditions from their homeland. Very early on it was decreed they could not bring their drums with them, nor participate in any ceremonies brought from their native land. They were treated as sub-human, much of that coming from the fear the white men who brought them here had and what they had heard of the Black cultures from which they came: tales of witchcraft and magic and worse. And the beat of the drums was seen not only as arousing animal passions, but also as a means of communication between people.

Once freed, which came very slowly after the end of the Civil War, the music of the African peoples began to develop its own rhythm and feeling, in what was now their land too. It started in the fields where they were forced to work, and the Christian Churches they attended, often by force too, and whose music they turned into their own. It developed into the blues and into jazz.

> Through music and rhythm carried by slaves and then by all those musicians and music and slaves, Africa has loosed itself upon the world. There is virtually no place that is not touched by its pulsing metaphysics incarnated in jazz, the blues and soul music, gospel singing, rock 'n'roll music that takes us south in our psyches. Hot

music loosens the religious corsets in which, until recently, most of the Western world has been so firmly encased. By changing the ways in which we move, dance, and sing, Mother Africa has given us access to spirit we never thought to have.

(Houston 1996 p. 57)

The great classic blues singers in the very beginning were women, like Ma Rainey and Bessie Smith. Bessie was a blues singer by the age of twenty. Those who heard her, first in the clubs, then on the many recordings she made, spoke of the elemental nature of her voice, which seemed to contain all of the terrible experience of the generations of Black people who preceded her in this land as slaves. Her songs were mostly of love, but love that was doomed, and they were about both men and women lovers, for Bessie chose whom she loved and lived at least that part of her life in freedom, with every part of her.

As we came into the 1920s, a Black middle class began to develop who were better educated and determined to blend into the majority white culture. And they were not fans of the blues, seeing them as vulgar and primitive, wanting to project a new image of a "whiter" Black person to their poorer, uneducated Black friends. The exception to this was the newer generation of art-minded Black people, such as Langston Hughes, who saw the blues as an essential part of the Black experience.

Bessie Smith died in the late 1930s, from injuries sustained in a car crash, and a newer form of the blues, mostly sung by Black men, came to the fore. It was not until a group of feminist scholars, in their research of the 1980s and '90s, came across the contributions that Black women had made to the beginning of the music form known as the blues and put them in their rightful place in American music history.

"For me, jazz has been a great epiphany of Dionysus in this century" (Lopez-Pedraza 1990 p. 41). In the very best singers and musicians that performed this music, a quality very like that of the best flamenco singers, dancers and musicians developed. A duende. Duende that rises from the feet and out of the mouth and musical instruments. A Dionysian cry from the very depths of the soul.

"Billie Holiday...she had a range of shattering verbal textures that are filled with duende" (Hirsch 2002 p. 46). She had that Dionysian cry. One of her most famous – or perhaps I should say infamous – songs was "Strange Fruit," a protest about lynching in the South. An executive in the Federal Bureau of Narcotics, who was a known racist, forbid her to sing it and she refused. At this point he began putting together a plan to destroy her. It was a couple of his agents who sold her heroin, and then she was caught using it and arrested. She served a year and a half in prison for it. Because of this, when released she could no longer get a cabaret singer's license. But she continued to sing in concerts at venues like Carnegie Hall and jazz festivals. And "Strange Fruit" was always part of it.

I heard her sing at the first Monterey Jazz Festival in 1958. I had seen her backstage, sitting not too far from where I was taking notes on what I had seen as an agent for the photographers there. She sat in her brocade cocktail dress, waiting to go onstage, and the slash of her dress from the hip to her knee revealed one of her thighs, covered with bruises. I thought they were from an abusive lover, and I was angry. They *were* from an abusive lover – heroin.

When signaled by the crew, she went on stage to thunderous applause from the audience. And when she sang, her voice breaking in places from long years of performing and deteriorating health, all the pain of her life, of her race and all it endured, came through. And the audience knew and felt that pain in her words and music, and experienced the duende. One year later, she was dead.

Note

1 "Lament for Ignacio Sanchez Mejias" by Federico Garcia Lorca, translated by Stephen Spender & J.L. Gili, from THE SELECTED POEMS OF FEDERICO GARCIA LORCA, copyright ©1955 by New Directions Publishing Corp. Reprinted by permission of New Directions Publishing Corp.

References

Cobb, Noel (1992) *Archetypal Imagination*. New York: Lindisfarne Press.

Goldberg, Susan (2018) For Decades, Our Coverage Was Racist. To Rise Above Our Past, We Must Acknowledge It. *National Geographic*, March 12.

Harrison, Jane (1992) *Prolegomena to the Study of Greek Religion*. Princeton, NJ: Princeton University Press.

Hirsch, Edward (2002) *The Demon and the Angel*. New York: Harcourt, Inc.

Houston, Jean (1996) *A Mythic Life*. New York: Harper Collins.

Lorca, Gabriel Garcia (1955 1998) *In Search of Duende*. New York: New Directions Bibelot.

Mitchell, Timothy (1994) *Flamenco Deep Song*. New Haven: Yale University Press.

Pedraza, Rafael Lopez (1990) *Dionysus in Exile*. Wilmette, Ill.: Chiron Publications.

Ponce, Charles (1980) *Lecture*. Santa Cruz, Ca.: University of California.

Shlain, Leonard (1998) *The Alphabet Versus the Goddess*. New York: Penguin/ Compass.

9 Dionysian women

We come into this world as Dionysian. Look at any infant blessed with well-functioning senses and it is obvious that the infant experiences the world through touch, taste, sight, sound and smell. And this remains its chief contact with the world until the world, through parents, teachers and other authority figures, begins to curtail this approach to life. For the most part this is conveyed through opinions given out that one sort of sensation is "not quite nice," but often it is through more stringent (and even physical) curtailment that is brought to bear. The rules for behavior vary with the culture into which the child is born, even the country of birth and often the religion of the family or nation. Behaviors thus curtailed usually make their way into the unconscious of the child and remain into adulthood, developing into what Jung called the "Shadow." He stressed that not only are negative tendencies stored here, but, along with them, much creativity. Fortunately, many of us are able to maintain our early relationship with the senses, and thus the Dionysian, but even then the society in which we grow up tends to punish us or judge us negatively if these behaviors are obvious.

Throughout the ages, there have been Dionysian women. Women who stand out and who have been the objects of both attraction and disapproval, often in projection. They live large and by their own rules. They are the later incarnations of the maenads. And since Dionysus' retinue included the Muses, very often such women were also in the arts. In fact, it is a blessing for a Dionysian woman to have an art into which she can let flow that terrific creative energy which soars through her. No longer can she run through the woods and up and down the mountains in ecstasy. Not in Western culture. So she pours that energy into what she creates. If she does not, then wine or drugs or other obsessions will overwhelm her and she will become merely the addict, bound for self-destruction. The modern Dionysian woman has always faced disapproval, laced with envy, as did the ancient Greek and Roman women who followed the God in ecstasy.

In Chapter 2 we highlighted three women to who Dionysus came in their older age to revivify their life and art. In the cases of O'Keeffe, Graham and Duras, their unique expression in the fields of painting, dance and literature were indicative of their connection to the Dionysian way of being. O'Keeffe's brilliant, in-your-face flowers and bones; Graham's ancient, dark-themed choreography set in even darker, stark settings; and Duras' novels and films exploring the deeper, darker side of human emotion, all bear witness to this.

Now we will examine three women of modern times whose private and public lives and art certainly showed them as devotees of Dionysus: Frida Kahlo, Josephine Baker and Colette, women of color, each from a different culture, but all maenadic in their very public lives and their art. And all three having a personal sense of drama in their professional and personal lives. Dionysus is God of the Theater.

Colette

Belle Époque, Fin de Siècle – it is no coincidence that these are both French phrases for a particular time in which art and sexuality flourished, for it was in Paris they reached their zenith. This was the historical period roughly drawn from 1871–1914.

It was during this time that several noteworthy events and ideas took root and became focused upon in the culture. In 1886, Richard von Krafft-Ebing, an Austro-German psychiatrist, wrote his groundbreaking book *Psychopathia Sexualis*, ostensibly for professionals in the field, but taken up avidly by European intellectuals. In it he wrote of female sexuality, necrophilia, homosexuality and incest, all of which were known but not spoken of, not written of, except by the notorious Marquis de Sade or Joris-Karl Huysmans in their novels, hailed by the intelligentsia of the times. It was also in this era that the wondrous Eiffel Tower was constructed as the entrance to the World's Fair in Paris in 1889. Romanticism, begun in the late eighteenth century, was slowly turning into Impressionism and writers and artists were gathering in Paris.

Into this world, in 1873, Sidonie-Gabrielle Colette was born, and within twenty years was living in Paris with her notorious first husband, Monsieur Willy, and becoming very much a part of this era, both personally and professionally. She was born in rural Burgundy and the daughter of a Dionysian woman, becoming one herself: "Sido extolled free love, expounded for her daughter an ethic as radical today as it was in Saint-Sauveur over a century ago, an ethic based on total freedom of the individual and the abolition of marriage" (Francis and Gontier 1999 p. xi).

Colette took her pen name from her father's last name. However, after a romantic tryst which rendered her ineligible for marriage as a dowerless girl, it was her beloved mother, Sido, who steered her into marriage at 21. The new groom was a 33-year-old writer and entrepreneur, Henri Gauthier-Villars, known as Monsieur Willy and already a celebrity in Paris.

Monsieur Willy and his new wife, who looked like a schoolgirl with the braid of hair she wore which reached to her ankles, became the toast of intellectual Paris – and more so when she wrote and published the sensual adventures of Claudine, a schoolgirl, though Monsieur Willy signed them as author. Everyone knew they were Colette's creation. Gradually, he encouraged her in lesbian affairs, sometimes a threesome. She had a penchant for theatricality and began to make appearances, first at private parties, then eventually on stage, as a dancer and actress. She was one of the first to dance nude on stage, scorning the gauze tights and see-through bodices worn by other performers and dancers. Only Isadora Duncan had performed in partial nudity before. When the half-drama, half-pantomime "Pan," written by the Belgian poet Charles van Lerberghe, opened, she was a natural for the role of Paniska, the maiden who has a liaison with the god Pan. Copying from Nietzsche, who blamed Christianity for the loss of all joy in sexuality and the senses, van Lerberghe's piece was a hit – as was Colette in it.

She went on to perform in other pieces, both pantomime and plays, and her relationship with the lesbian Marquise de Mornay culminated with stage appearances by them both, creating great scandal in Paris. The Marquise dressed as a fashionable, middle-aged man both in private and onstage. Colette moved in with her, or at least took an apartment behind hers, which she used merely as a garconniere. Her relationship with Willy continued in a way that confused her lover, her mother and those around her. They no longer shared quarters, even went through a division of furniture and belongings, but she still could not seem to let go of him, even as he entered live-in relationships with other women. Finally they did divorce, and this freed Colette to enter into another sort of life, though slowly.

Her threesomes and lesbian alliances continued until the Belle Époque itself came to an end in 1913, both for Paris and for Colette. She was forty years old, and her pregnancy and subsequent marriage to Baron Henri de Jouvenel started a new epoch in her life:

> Sido and the Belle Époque died together. In 1913, the year that
> a motherless and pregnant Colette would embark on a new life,

Diaghilev produced Stravinsky's *Rite of Spring* with choreography
by Nijinsky. Alain-Fournier published *Le Grande Meaulnes*...and
Proust, *Swann's Way*.

(Thurman 1999 p. 243)

Colette remained a Dionysian woman, but failed utterly in the role
of mother, to the point where she seduced her teenage step-son by
Jouvenel and entered into a five-year liaison with him. Her daughter
by Jouvenel, Bel-Gazou, was for the most part ignored by Colette and
raised by governesses until departing for boarding school at age eight.
Her marriage once again ended in divorce, with Jouvenel's son taking
his step-mother's side in the proceedings. The writing continued, however, and Colette's fame as a writer endured throughout all the scandals.

As Paris and the world began to show interest in Africa and its art,
she did an interview in Nice in which she spoke of her Black ancestors
(her great-grandfather on Sido's side), putting herself right at the
avant-garde of the new era. In 1925, she met the Dutch-born dealer in
diamonds and pearls, Maurice Goudeket, fifteen years younger than the
then 50-year-old Colette. "He encouraged her to quit journalism and
to write what have become known as her major works: *The Break of
Day*...*Sido*, and *The Pure and the Impure*" (Francis and Gontier 1999
p. x). Goudeket was the last of her three husbands and stayed with her
until her death.

Her love of him was to bring a fear and anguish not known to her
before, as during World War II, when the Germans marched into Paris
and took over, Goudeket, being a Jew, was put in a concentration camp
outside Paris and kept there in preparation to being sent to Auschwitz.
Colette pulled all the strings she possibly could with friends still influ-
ential until he was freed, though he had to live in hiding until the Nazis
were driven out. This took most of her strength, though she lived on
another ten years, to 1954. But arthritis took its toll on her arms as well
as much of her body, and she was ultimately unable to write.

When she died, she was the first woman writer to be buried with full
military honors in France. She is buried, as are so many other famous
people, in Pere Lachaise Cemetery.

Josephine Baker

The Paris into which Josephine Baker stepped in 1925 was already
a haven for African American artists, many of whom discovered it
in World War I as soldiers and servants in the American Army. The

French gave them a very different reception to what they were used to in their own American culture – no segregation or the daily humiliations experienced back home. The French loved their style and their jazz, and welcomed them into the fold.

However, it was the appearance of Josephine Baker in La Revue Negre, with her explosive dances, that focused the attention on African artists and culture worldwide. An American ex-pat writer living in Paris at that time, said it all: "the place was set on fire by the most sensational woman anybody ever saw. Or ever will. Tall coffee skin, ebony eyes, legs of paradise and a smile to end all smiles" – this from Ernest Hemingway after seeing Baker's show at the Theatre des Champs Elyse (Lahs-Gonzales 2006 front cover).

This echoed her reception by the French: "The highlight of the show was a wildly erotic dance, referred to in the press as a Dionysian spectacle, performed by Baker" (Lahs-Gonzalez 2006 p. 9).

France, and the world in general, had been experiencing an African genesis for some time. It had begun in the nineteenth century with the colonization of much of Africa. Colonists and soldiers returning home had brought a great many "curios" from the assorted countries they had conquered and colonized. These were mostly seen as "primitive," sometimes in an effort to discount their importance.

In the early twentieth century, however, it was the art world that began to see the value in these pieces, albeit while still having little understanding of the complex cultures behind them.

The Paris Exposition of 1900 was the first major display of African art, "an explosion that tore through the fabric of Western Christian artistic sensibility that had its hold on the world for the past two thousand years" (Anthony 2007).

Seven years later, a young Pablo Picasso came across an African exhibit at the Muse d'Ethnographie in Paris:

> From this encounter, from even this very moment, for he uses the word "suddenly", he understands why he is a painter. The masks he is looking into, he realizes are magical. They possess real power. They are weapons to use against the dangers of life.
>
> (Griffin 1992 p. 213)

And not just Picasso. Many other major painters of the School of Paris, Matisse among them, were finding magic and meaning in these African "curios" (still also being sold as such in junk shops in Paris) and seeing them as art. These artists began to use this African influence in their own work, Picasso's *Les Demoiselles d'Avignon* being a case in point, with its five prostitutes, two of them wearing African-looking masks, which changed the face of the Paris art world. German Expressionist painters caught the "primitive" fever and a congruent Polynesian flavor too, through a 1910 Gauguin exhibition. "Primitivism" was in. And also "Americanismus," which was fostered by the African American musicians whose music dominated the cabaret scene in Berlin throughout the 1920s, playing the jazz that Berliners seemingly could not get enough of, and which was regarded as "jungle rhythms in which a mythic Black sexuality and its expression was seen as a refutation of established bourgeois value" (Garretson 2016 p. 11). In still-segregated America, at the prime of the Harlem Renaissance, a Black philosopher, Alain Locke, was directing African American artists to look to Africa for their inspiration.

The drive to the so-called primitive, which I believe to be a return to the Dionysian way of life, also called out a variety of European intellectuals and writers, C.G. Jung among them. He traveled to Africa in 1920, then again for a longer period in 1925 to Kenya and Uganda. In his quasi-autobiographical *Memories, Dreams, Reflections* (quasi because of recognition in recent years of its being greatly edited by Aniela Jaffe, to whom it was dictated, and of material excluded by the Jung family), he writes of the impact of the trip on him: Africa, "of primordial beginnings...maternal mysteries, this primordial darkness" (Jung 1992 p. 269).

This was the Europe into which Josephine Baker, an 18-year-old African American girl from St. Louis, Missouri, stepped, and immediately not only became part of this scene, but one of its focuses. Writers such as Gertrude Stein, Simone de Beauvoir, Ernest Hemingway, Langton Hughes and e.e. cummings paid instant notice by writing about her. Architects Adolf Loos and Le Corbusier met her and wanted to design for her. Artist Alexander Calder did five of his wire and iron sculptures of her, and Kees Van Dongen painted iconic images of her and more. Even the most famous couturiers of the time – Paul Poiret and Jean Patou, amongst others – competed to dress her. Baker became what Le Corbusier, in his book *Precisions*, identified with admiration as The Modern Woman.

She came from a very poor family in St. Louis, Missouri, and her mother kept her paternity in doubt, though there are indications her father was white. Partly this was because of her being born in a hospital and having a birth certificate, which was not often the case for Black Americans at the time, in 1906. From the beginning she was to show that she was her own person, having little time for school, and hanging around outside theaters where she would show the performers and producers as they went in that she could dance, could sing, could do whatever they wanted so she could get on that stage. Eventually this led to her doing bits in shows, which in turn led to New York, then to Paris and the Revue Negre. From the moment she danced onto that stage in her famed banana skirt, doing what she later confessed was a completely improvised dance, she was a hit. She used her body as her canvas, her art form. Dionysus is the body. She used it to dance – dance salaciously at first, to provide her Paris audiences with their projected image of the Dark Continent, which both enticed them and frightened them. A year later she was dancing in Berlin, and was feted there also, though afterwards she was to say that she felt an undercurrent of prejudice in Vienna. Gradually, still using her body, she draped herself in sensuous, glittering, see-through fabrics which revealed more than they concealed.

However, as Andre Rouverge noted in Mercure de France, "The girl has the genius to let the body make fun of itself...To be sure, her body shakes as if in a trance, but with remarkable humor." And Michael Borshuk "argues that the exaggerated characterizations of the 'savage' were in fact a subversion of the black stereotypes imposed on her" (cited in Lahs-Gonzales 2006 p. 45). "Baker's body, which was athletic and almost boyish, enhanced her androgyny, which she sometimes exploited. 'Josephine Baker. Is she a man? Is she a woman?...She is horrible, she is ravishing, is she black?...Is she white?'" (Lahs-Gonzales 2006 p. 45).

And, of course, androgyny was always part of the Dionysian story.

At a certain point she capitalized on her Parisian popularity by going into the cosmetic industry (much like her friend Colette would do), producing a skin-darkening lotion she named Bakerskin and a hair pomade named Bakerfix, for fans anxious to duplicate her look. The profits from this business would enable her to buy a château in Southwestern France, which she dubbed "Château des Milandes."

She continued her career in Paris, though the famed "banana skirt" was left behind as her costumes became more elegant. At one point she went back to New York and performed in the Ziegfeld Follies, but ran into the same old American racial prejudices, and came back to Paris, to stay. She married a Frenchman, gaining French citizenship.

When World War II began, she immediately offered her services to the Free French, and the French Underground recruited her to carry messages to the different European capitals where she was still performing and doing shows. She carried these in her music, and sometimes on her person in her costumes. Dionysus would have loved the drama of it all, and the closeness of death if she had been caught.

In 1947, she began to fill the twenty-four rooms of her Château des Milandes with adopted children of various nationalities, religions and colors. Her Rainbow Tribe, she called it. She was married at the time, and eventually wrote a children's book about the children and herself called *La Tribu: Arc-En-Ciel*. She battled for many years to keep the château, but finances failed and she eventually lost it.

Throughout, she remained outspoken about racial prejudice, her beloved Rainbow Tribe giving a positive slant to her feeling that people of all colors, religions and races could live peacefully together. This was recognized finally in the United States when, on August 28, 1963, she was the only official female speaker at the March on Washington, where Martin Luther King gave his famous "I have a dream…" speech. It has been written that after Dr. King's assassination, Mrs. Coretta King asked Josephine to return to the United States to lead the Civil Rights Movement. But the pull of her twelve adopted children was too great, so she declined and remained in France.

In her sixties, she took to the stage again, one such show being a revue, "Bobino." Her costumes were designs by Dior, Balmain, Balenciaga and other fashion greats. At one concert, in 1963, she opened her show by saying "Not bad for sixty, huh?," to thunderous applause.

The French government awarded her the Croix de Guerre, the Medaille de la Resistance and the Legion d'Honneur, in recognition of her services in World War II. She died on April 12, 1976, and more than 20,000 people lined the streets of Paris, where she was given a twenty-one-gun salute and was the first American woman to be buried in France with full military honors. Not bad for a poor, Black girl from St. Louis who traveled far from home to do what she felt called to do as a Dionysian woman.

Frida Kahlo

> Frida Kahlo was more like a broken Cleopatra, hiding her tortured body, her shriveled leg, her broken foot, her orthopedic corsets, under the spectacular finery of the peasant women of Mexico… The laces, the ribbons, the skirts, the rustling petticoats, the braids, the moonlike headdresses opening up her face like the wings of a dark butterfly: Frida Kahlo, showing us all that suffering could not wither, nor sickness stale, her infinite variety.
>
> (Fuentes 1995 p. 8)

In reading Fuentes' description of her, I was moved to think that Frida Kahlo is a metaphor for all women – the bangles, glitter and color

covering the wounds, the humiliations in each of us that the patriarchy imposes on women (and many men) and which we have historically sought to disguise and perhaps forget. The Dionysian woman, like
Frida, does this with flair, flaunting the mores that attempt to restrict
them, breaking those barriers through their art. It could be said that
Frida WAS her art. Most often it was her own face she painted, with
a nakedness that is very striking, and sometimes shaking to the viewer.
The feelings are there, laid bare.

Though she was actually born in 1907, Frida would consistently
claim it was 1910, the year of the Mexican Revolution that lasted for
ten years. Thus, she identified not only politically, but as a revolutionary
by her very nature. This showed early in her school days at the Escuela
Nacional Preparatoria, where she intended to study medicine, which
was then a predominately male field of study, and by her association
with a predominantly male group calling itself the Cachuchas, who met
to discuss politics and academic matters. She also indulged in other,
smaller transgressions, such as wearing bobby socks (which were forbidden by the dress code) and publicly disclosing her intimate relationship with an older woman – this, in addition to her intimate friendship
with a male member of the Cachuchas. All of this, however, was cut
short by the terrible, gruesome injuries she sustained in 1925, in a bus
collision, when only eighteen years old.

Even with the advances in medicine and health care today, more
than ninety years later, the severity of her injuries – including multiple
fractures, a crushed foot and the piercing of her body by a pole in the
bus – would be a miracle of survival. However, survive she did, and she
spent a year in bed following the accident. It was at this point, with
an easel designed especially to be used by her lying in bed and with
brushes and paints, that her destiny as a painter began to unfold. At the
end of that year, a few canvases in hand, Frida made her way to where
Diego Rivera was working on yet another of the murals for which he
was famous, called up to him as he was standing high on a scaffolding,
painting, and asked him to come down and speak to her. Perhaps
astonished by the audacity of such a request from an unknown, though
pretty woman, Rivera descended. She asked straight out that he give
her an opinion on her work and whether it was worth pursuing in order
that she might help with the expenses incurred by her family for the
treatment of her terrible injuries. She gave him her address and asked
that he visit her at home in a week and give her his verdict. He came,
setting in motion the two obsessions of her life: painting and Diego.

Much has been written in the biographies of Kahlo of the portrait she
painted of the two of them a couple of years after their marriage – the

marriage characterized by her father as the wedding of an elephant and an a dove. Most concentrate on what they see to be the disparity of the two: Diego, huge, palette in hand, announcing he is an artist, looking away from Frida; Kahlo, looking like a traditional Mexican wife of the time, small next to his stature and girth, her hand on his. Few have seen beyond this. However,

> The inscription on the ribbon along the top of the composition begins with a seemingly benign identification, "Aquinos veis, a mi Frieda Kahlo, junto con mi amado esposo" (Here you see us, me Frieda Kahlo, with my beloved husband Diego Rivera) which prompts reading the painting as a marriage. But the subsequent statement, beginning with "Pinté estos retratos" ("I painted these portraits"), emphasizes Kahlo as producer, clarifying that while "here you see us", it is "me Frida Kahlo" who has created this double portrait.
> (Lindauer 1999 p. 18)

The author goes on to point out that the hands barely touch each other, and that the red shawl, the red in the flowers on her shoes and the red dots on the ribbons in her hair dominate in the otherwise subdued colors on the canvas. This was the Frida whose father warned Diego at the marriage, "She is the devil."

I believe Frida was her own person from her youngest years. That horrific bus accident only served to put her more deeply into herself, and the mirror placed in the canopy over the bed in which she recovered at home came to mean more than simply depicting herself to herself as subject matter. As the years went on, both she and Diego came to collect pre-Columbian art. Surely in that collection there were images of one of the most powerful gods of that time, Tezcatlipoca, known as Smoking Mirror. This was one of His names as his right foot, which had been crushed in battle, was often replaced in depictions of that foot as a mirror, a smoking mirror, related to the obsidian used as mirrors in shamanic rituals and for prophecy. Obviously Frida, whose own foot had been crushed and whose bed mirror metamorphosed her into the self-portraits for which she is most famous, felt a link to Him. Like Dionysus, He was a deity of the Night, and, like Him, had a brother who was His polar opposite: Quetzalcoatl. Tezcatlipoca, too, was often portrayed as a youth. "Tezcatlipoca represents the equivalent of what might be called the 'shadow' – the side of our human personality we do not wish to face openly, and which we consequently hide from ourselves" (Burland and Forman 1975 p. 56). Sounds quite the Dionysian, which has been repressed for two thousand years all over European civilization.

I am not an art critic, therefore I believe what I respond to in Frida's art is what masses of other women continue to respond to: the bare, female truth that stares through her self-portraits and all her work – as in the masks of Dionysus that were mounted on Greek theater walls, the face usually staring straight ahead, looking directly into us, daring us to look deeper into it.

The first real wave of Frida awareness came in 1979 with the publication of Hayden Herrera's (1983) biography of her. Women took Frida up as a heroine and she became a cultural phenomenon, and remained so with exhibitions of her work popping up everywhere in museums, as well as exhibitions of the work of young Mexican artists, prompted by her popularity, drawing on her for inspiration. I went to several of these myself and bought many of the seemingly endless books about her that followed. Always it seems that her image, her style, her life was part of this tide of popularity, in addition to her art. Most recently, in the late Spring of 2018, it was a show at the Victoria and Albert Museum in London, "Frida Kahlo: Making Herself Up," which drew, along with massive crowds, some critical condemnation. Most of the objections concerned the presence of her prosthetic leg, which she had to wear much later in her life when her badly damaged leg had to be surgically removed. Her defiance led to her wearing a bright red boot on the foot of it. Critics objected to the leg, with its bright boot, being placed in a display case. Yet its very inclusion speaks to the flamboyance and defiance of convention with which Frida lived: "Making Herself Up."

Frida loved retablos placed by families in shrines and churches asking for a miracle for the scene painted on a piece of metal or wood, representing what needed healing. Also, it was not uncommon to find discarded crutches and canes and other medical devices left there by the grateful victim or her/his family. Frida herself left no painting of her accident, except to adapt a retablo she found of a bus accident, adding her female figure on the ground nearby. This is true in New Mexico in this country, in such places as Chimayo at the shrine of the Santo Nino, and other churches and shrines, where items are left by cured folks who visited the shrines and prayed for healing. However, I believe it is more than that, too. I believe that the very Dionysian quality of Kahlo's life and art attracts the new wave of feminism happening now, and for the last few years. Her art is powerful, and it is important to note that it came from a woman, and a woman who very much was her own person, who overcame odds to become that painter and that person, with difficulties that would have defeated many another person, let alone a twentieth century woman. That these odds are concurrently on display with her art seems fitting. I would, however, concur with the writer that more of her artwork would have been good, to balance the many personal items on display in the exhibition.

At a later date, a woman researching Frida's choice of Tehuan folk dress felt that it was Frida's way of proclaiming the matriarchal society of the Tehuan women who were in charge, on the practical level of the needs of their society.

Feminists in general protest when the art by women is touted as "women's art," yet until societal prejudices die down – or, preferably, die out – it remains necessary, I believe, to announce that "this is art by a woman," art every bit as strong and creative as that of any male artist. I don't think that women's art should refer to an art that is especially feminine, which it seldom is, but to draw attention to the fact that there are so many women out there creating art that is as effective as any other art of this time, and for the many centuries before.

So Frida Kahlo, living as the equivalent of a maenad, with all the creative energy from that life flowing into her art, stands, I believe as a great role model for us.

Her life has been so thoroughly documented these past forty years since the publishing of Herrera's book: the obsession and disappointments of her love for the womanizing Rivera, her constant pain from multiple injuries, then multiple surgeries to try to fix them, which I've not dwelt on here. Her art and her life contain the Dionysian cry perhaps louder than any words could.

No women from the late twentieth or early twenty-first centuries have been included here. I know they are out there. However, in this age of celebrity worship and clamor, and of selfies wildly distributed on the Internet, it is hard to discern between natural Dionysian women such as the three I've written of in this chapter, and those who behave outrageously in pursuit of the media hype and the fascination in our culture for those who are famous simply for being famous or because they were featured on television reality shows. To behave wildly, to drink or drug to excess, to act extravagantly does not necessarily indicate this is a true part of the personality, but rather a way to attract the attention of the world. Dionysian, perhaps, but not coming from a natural conviction as with the women of whom I've written. There might have been a narcissistic element to those women, of course, but their behavior and attitudes seem to have come from within, rather than being manufactured to get attention. And their drive manifested in their artistic work.

I know there are women out there today who are true Dionysians, but I have chosen one woman whom I feel has achieved balance in this century and who is making the attempt to summon this latent power in other women in this century, and the years to come, where she feels the danger is great for our annihilation of humanity. And it is this woman I write of next.

References

Anthony, Maggy (2007) *Critical Mask Mythic*. Imagination Institute online. www.imagination-institute.org/

Burland, Cottie Arthur and Forman, Werner (1975) *Feathered Serpent and Smoking Mirror*. New York: Orbis Publishing.

Francis, Claude and Gontier, Fernande (1999) *Creating Colette Vol. 1&2*. South Royalton, Vermont: Steerforth Press.

Fuentes, Carlos (1995) Introduction. In Kahlo, Frida, *The Diary of Frida Kahlo*. New York: Abrams.

Garretson, Tom (2016) *Dancing at the Edge of the Abyss*. Master's Thesis. Department of Art History, University of Oslo.

Griffin, Susan (1992) *A Chorus of Stones*. New York: Doubleday Books.

Herrera, Hayden (1983) *Frida: A Biography of Frida Kahlo*. New York: Harper & Row.

Jung, C.G. (1992) *Memories, Dreams, Reflections*. New York: Vintage Books/ Random House.

Lahs-Gonzales, Olivia, ed. (2006) *Josephine Baker: Image & Icon*. St. Louis, Mo.: Reedy Press.

Lindauer, Margaret A. (1999) Devouring Frida: The Art History and Popular Celebrity of Frida Kahlo. University Press of New England.

Thurman, Judith (1999) *Secrets of the Flesh: A Life of Colette*. New York: Alfred Knopf.

10 Jean Houston's mythic life

After reading a few of her books and watching some of her YouTube videos on my computer, I have become convinced that the message Jean Houston carries is a most useful one for women today and in the future. I would like to begin with some background to my thinking on this.

In 1973, I attended a weekend seminar with Joseph Campbell as the featured and only speaker. The subject matter was "The Hero's Journey," a subject which, as both a fan of the Arthurian story and a student of Jungian psychology was of great interest to me. From the introductory evening through to the last day's presentation, I was very impressed by the fact that Campbell used no notes in speaking of his subject, a sign to me that he was thoroughly immersed in it after years of study.

On the final day, after he had wrapped up his talk, he asked if there were any questions from the huge audience. My arm shot up almost without my knowledge, and to my surprise I was called upon by him immediately. My question was ready: "You've spoken to us these last days about the hero's journey, but what of the heroine's journey, the woman's journey?" He gave a great sigh as he looked out at the audience, which I instantly interpreted as a "oh, no! a feminist!" He proceeded to say we had no notion of the woman's journey as no woman had ever written about it. I sat back down, stunned at his reply, and he went on immediately to answer another question. Going through my mind were names of women who HAD written of their journeys: Colette, Anais Nin, Emily Dickinson…and many, many others, through the last hundred or so years.

This occurred in 1973, and still rankles, since I remember it so vividly. So, imagine my delight when listening to a lecture by Jean Houston this year, some forty-six years later: at the beginning of that lecture she claimed we are now living in a most critical time, with the need to achieve a new humanity; she went on to proclaim that she felt it was woman's work to achieve this. She then brought up her friendship and association with Joseph Campbell, and spoke of their only real battle being her insistence that a woman's journey was different than a man's; she went on to speak of those differences. I felt a kinship with this woman at that moment, and a more complete understanding of Campbell's reaction to my question. With all his brilliance and scholarship, he was still a man of his era, where women were at most second-class citizens.

In that lecture, she went on to delineate the differences in the hero's journey, a refined product of the 2000-year-old mindset about the role of the man, with the role of woman. What I got was that the role of women was more one of inner battles and struggles, as opposed to the outer battles of the hero, solved with knives and lances.

All this led me to buy a copy of Houston's *A Mythic Life*, which is a mix of autobiography, thoughts and philosophies she developed over the years from her life experiences. One paragraph stood out almost immediately to me:

> Consider my fantasy of the possible human, a once and future person, who may be both what we were and what we may yet become. The first thing you notice about her is that she enjoys being in her body. A fullness of being inhabits that body, with its flexible joints and muscles, its movements fluid and full of grace. One senses an ebullience in the bones, an appetite for delight. She is given to long pleasures and short pains...
>
> (Houston 1998 p. 40)

This sounds very Dionysian to me, though she makes no reference to him and there is no reference to him in that book. However, we are once again in the area of the deep feminine to which Dionysus is linked.

As Jean Houston stated in her lecture, I believe it is women and the values of the feminine that can do the most to save our civilization from the dark path it is now on.

She has lived a long (eighty-two years!), active life, both intellectually and physically.

She worked as a consultant to the United Nations, visiting more than 105 countries. She has delivered countless seminars and classes, and is now specializing in mentoring classes which span a period of months, after the initial seminar in her mountain top house in Ashland, Oregon, following up with monthly teleconference calls. It is called "I Am Woman Hear Me Roar." In these classes she gives the benefit of her more than six decades years of experience out in the world. And speaks of moving together as cells of a living body of womankind. A helpful revolution indeed.

Houston concentrates her training seminars in quantum physics theory. She works with quantum powers, which she (as do others) believes create a partnership with what she terms the creative source. This, she believes, will benefit not only one's own life, but all around one and the emerging collective future which is desperately in need of help at this time.

She has had strong contact with some of the most intelligent people of the twentieth century, and therefore I feel she is in a position to see a way forward that will benefit humanity and save us from itself.

Reference

Houston, Jean (1998) *A Mythic Life*. New York: Harper One.

11 Dionysus appears, again

From all the stories told to me about my earliest childhood, I can identify myself as a Dionysian, like all children. That all seemed to come to an end when I developed allergic asthma, and then chronic tonsillitis. I was often sick and out of school (a plus for me, as school was difficult with all the bullying I endured, which was perhaps a contributing factor for my sicknesses). So, a great part of my childhood was spent immersed in books and daydreaming.

In my early twenties, I had decided what I most wanted from life was adventure. I met and married a man who loved adventure too and who was then an up-and-coming photojournalist. In those days magazines ruled the information highway, along with newspapers and television. When the opportunity came to go to Brazil, at the invitation of some young Brazilian filmmakers we met at the San Francisco Film Festival in 1962, we packed up and stored our belongings. My husband got an assignment from *Playboy Magazine* for a "Girls of Rio" spread, and from *Life/Time* a job as a stringer photographer in their Rio office.

My life until then had been largely lived in my head and my fantasies, my body almost non-existent for me. I didn't dance, didn't do any form of exercise and largely never thought about it. As a plump, though not obese, child I was mercilessly teased at school and barely noticed it when, at about fifteen years old, I was suddenly a slender young woman, just as I entered high school. It didn't seem to make much difference except that I was now receiving attention from boys. That didn't make much difference either. I loved books and writing and classes in which these featured, and my life was largely lived in them.

With my arrival in Rio de Janeiro, at carnival time in 1963, all my long-dormant senses awakened, and with them some fears about these hitherto unlived feelings. I was about to experience Brazil as the Great Awakener for me.

The aroma as I exited the ship in the harbor at Rio de Janeiro! The docks smelled of what I was to learn was rotted sugar cane, and, further into the city, the smell of "parfume" that permeated the ongoing carnival celebrations. It was actually ether in spray cans sold on the streets. And then, once in any of the buildings, there was the smell of mildew from the intense humidity that comes with the summer.

That evening, after we had put our belongings in our hotel room and gone out onto the streets, the rhythm of the samba beat was felt all over. Music would be heard at a distance, then grow closer as I walked toward it. Finally, I would turn a corner and there would be a small band strolling back and forth playing the music of the carnival. Here and there, people would set down their packages and begin to dance the almost ritualistic dance steps peculiar to carnival. The following carnival, I was one of the Cariocas who set down her packages and danced. But it was a long and confusing year that I was to spend, not realizing that Dionysus had once more come to claim me as one of his own.

My body awakened, first to the sunlight on Copacabana Beach. Because I was so fair – blonde, blue eyed and pale skinned – I had been told I couldn't tan. Mornings on the beach in front of our hotel and suntan oil proved them wrong. I became a soft tone of bronze all over. With the heat and humidity and sensuousness of the city, I became aware of my body twenty-four hours a day. And every night, we went with a group of mostly Brazilian friends to dance at one of several boîtes in the city. The Black Horse was a favorite, and often I would dance until the wee hours of the morning. There, with one of the film stars we had met at the film festival in San Francisco, Jece Valadão, I met an Italian Countess, quite rich, quite beautiful, who offered to take us to an Umbanda ceremony.

Umbanda is the religion introduced to Brazil by the Africans who were brought there as slaves in the eighteenth century. It permeates Brazilian culture, in which the Roman Catholic Church is the biggest presence. As a friend in Rio said to me, "I am Catholic, but I have SEEN Umbanda." Once the religion was there it mixed with Spiritism, but remained essentially African. To satisfy their white masters, they accepted the saints and the Son and Mother of Christianity, because they could see similarities to their own "orishas" or emanations of the Divine. Thus, the Virgin Mary was identified with Iemanja, the Mother Goddess, St. George (a favorite) with Xango, and on and on. The Church was satisfied, and so were the people who were enslaved but could, at least, keep their own religion.

We scheduled a meeting in which the Contessa and her husband would take us to one of their midnight ceremonies along with a couple of Brazilian friends. I have written of my experience there in my book on the *Jungian Women* (Anthony 2017 pp. 53–54). And, as I have written, my experience there confused me even more. It was the drums, the chanting, the burning candles and the immediacy of the religious experience of the mediums that captured me, even as I couldn't quite understand it.

At one point, one of their major white mediums wanted me to join his psychic circle and go into trance for the then President of Brazil. As I have an innate fear of going into trance or being hypnotized, I backed off immediately. President João Goulart was a great believer, evidently, and consulted them often.

And while the practices were also popular among some foreign visitors, for me it was a very deep, personal Dionysian experience, one from the very depths of my being. It wasn't until thirty years later that a book by a Jungian woman analyst came out that made that part of my experience that night clearer: *Women Who Run with the Wolves*, by Clarissa Pinkola-Estes. In being "chosen" that night by Iemanja, I had made contact with that archetypal female energy within myself. It had always been there, in my psyche, but the culture and the times in which I was raised did not acknowledge it. As Dr. Pinkola-Estes pointed out, She has many names in many cultures: La Loba, She Who Knows, Wild Woman. But in the middle-American culture of the 1950s and early '60s, there was no speaking of Her, except in stories, and most often in a pejorative way. But She had always been lurking inside me, evidently, as years later, attending the wake of my 95-year-old cousin who had recently passed, one of the older women acquaintances of hers, when I introduced myself, said "Oh yes! You were the wild one."

Yes, the wild one who had not settled down and raised a family while in my twenties, but instead had traveled and sought adventures in Paris, Rio de Janeiro and Zurich.

> Old One, the One Who Knows, is within us. She thrives in the deepest soul-psyche of women, the ancient and wild Self…It is this point where the I and the Thou kiss, the place where, in all spirit, women run with the wolves.
>
> (Pinkola-Estes 1996 p. 830)

I am convinced that it is contact with this archetype in the deepest realms of the unconscious that will help women in remaking the goals of humanity to save our civilization.

I bring the deep feminine in as it is important to remember that Dionysus was connected from the very beginning with what was originally thought of as the Great Goddess. This is perhaps why as he disappeared outwardly in our society, the position of women did too, from the important position they had maintained since the earliest times prior to the takeover of the patriarchal forces that have governed since.

This was before Jung and his psychology came into my life and I learned my psychological type was that of Intuition/Feeling with the Sensation function being totally unconscious. What this meant for me was that all the material of the senses and sensations was unknown to me consciously, so was very powerful when experienced. This affected my whole experience of this country that thrives on color and music and movement. Finally, it was while in Brazil that my brother in the United States sent me *Memories, Dreams, Reflections* (Jung 1992), which changed my life dramatically. It brought me the intellectual understanding that I had to have in the midst of this dramatic personal experience.

In the almost two years we lived there, the Dionysian side ruled. Not only my life, but my early attempts at writing. Even my one experience of being in a movie smacked of the senses as I only danced and appeared drugged in the one scene I was in. Being a "movie star" had been a childhood dream, but that meager experience convinced me that wasn't my path. But, when time came to leave, after the revolution, I wanted to keep that experience and those feelings. I remember saying to myself "I must keep dancing, even when back in the United States." But, of course, I forgot, and it was several years before Dionysus tried again and I was able to make contact with the wild inside me.

I've come to find that many women, once I've shared my experience, have had similar awakenings in their lives, triggered by an unexpected event – sometimes falling in love with someone, often in inconvenient or difficult circumstances. And, as I have written in a previous chapter, in older age He comes in more helpful ways.

And it is in eras, too, that the Dionysian will manifest itself. The San Francisco to which I returned in 1964 was on the brink of its own Dionysian wave, which broke over the whole city, but in particular one specific neighborhood: Haight-Ashbury, soon to become home to the now historic Flower Children and the beginnings of a third wave of feminism, the Black Power movement and other social revolutions. With my husband, a photojournalist, it became up close and personal, as well as a worldwide phenomenon.

References

Anthony, Maggy (2017) *Salome's Embrace: The Jungian Women*. London: Routledge.

Jung, C.G. (1992) *Memories, Dreams, Reflections*. New York: Vintage Books/ Random House.

Pinkola-Estes, Clarissa (1996) *Women Who Run With the Wolves*. New York: Ballantine.

12 Dionysus rampant

Haight-Ashbury at its highest

The era of the Flower Children, or "hippies," as they were often called, actually had its beginnings in the late 1950s in San Francisco's North Beach and the Beat Generation. I had been there for a year in 1958, living and working with a photographer, Jerry Stoll, who was working on a book idea which eventually became a photographic volume entitled *I AM A LOVER*, the title being taken from a Walt Whitman line of poetry: "I am a lover and have not yet found my thing to love." Jack Kerouac and Allen Ginsburg had recently left the scene, but their influence remained. My days and nights were spent there, getting to know the scene, making friends with the poet Bob Kaufman, drinking wine and listening to poetry at the Co-Existence Bagel Shop and Miss Smith's Tea Room and other locales while Jerry took his pictures. It was a racial mix, unusual even in San Francisco, and Kaufman was hassled a lot by the police because of the white women who gathered around him, including his wife.

In a protest against the multiple tour buses which jammed the one-way Grant Avenue, filled with tourists gaping at all of us like watching animals at the zoo, I rented a tour bus and organized The Beatnik Tour of the Capitalist Wasteland. In a bus filled with Beatniks, we went to downtown San Francisco, to its poshest hotel at that time, the St. Francis, and wandered through the lobby with one of the young men using a megaphone to describe the businessmen and others found there, and then over to I. Magnin's very upscale clothing store, where we did an impromptu fashion show of the young women among us, with their black tights and very casual clothing, looking like the Parisian bohemian existential community much admired by all. We ended with a baseball game of the police versus the Beatniks, and, as usual, the cops won. All in good fun.

Evenings sometimes were spent at Eric Nord's Party Pad, where mostly African American young men would beat their conga drums while we danced. Yes, there was marijuana, and wine, but I never saw nor was offered anything stronger. Dionysus was present in His most congenial form, with poetry, dancing, jazz, free love and endless conversations. A comparatively mild beginning to what became, just a few short years later, the Haight-Ashbury scene, which also started out mildly.

The neighborhood of the Beatnik scene was what was called North Beach, which for years had housed a large Italian community bordering on Chinatown; Broadway Street ran through it, with its nightclubs and restaurants. Haight-Ashbury was a family neighborhood, filled with Victorian houses, and Haight Street, with its shops, and a few restaurants and bookstores. It was low-rent at that time, and that was a draw for the young folks who began to gather there. This generation appeared to have more of a social awareness, and before long a Free Clinic was opened to give medical treatment, and The Diggers organized to give free food. Racially, it was even more of a mix than the Beat scene. A couple of years earlier, Betty Friedan's *The Feminine Mystique* had come out, and a second wave of feminism began. The Black Power and Civil Rights movements were in full swing. And, of course, the press came and the publicity went out, and before long the word "Hippy" began to be used to identify the hundreds of (mostly) kids coming in. Any teenager who disappeared anywhere in the country was eventually sought in the Haight. It was much more of a revolution than the Beat scene ever had been, and this attracted many more young people, the 1950s middle-class lifestyles of their parents seeming pale and rigid by comparison. Of course, rock music was the dominant sound. It spread all over the world. While I had been in Brazil in the early '60s, The Beatles, The Rolling Stones and several other (mostly English) musical groups came to dominate the music scene. Another favorite was The Doors, whose lead singer, Jim Morrison, even had a widely circulated picture of himself as Dionysus, a wreath on his long, curly hair. I heard them, first in Brazil and then in the Haight, sprinkled in amidst Chubby Checker and the Twist and the samba, but not to the extent they were now dominating the record industry, television and concert halls. And before long, in the Haight, local musicians, often sharing a house, were beginning to make themselves heard too: The Grateful Dead, Jefferson Airplane. Early on, Bill Graham came into the scene; he took a defunct hall in the nearby Fillmore District. On December 10, 1965, he opened the Fillmore Auditorium with Jefferson Airplane playing and set a new precedent for rock concerts.

Nearby, bordering the Haight, was the Golden Gate Park, a beautiful swath of green trees and land which stretched all the way to the Ocean

Beach. And the hippies began to gather there too, for concerts, and to listen to some of the spokespeople who appeared, poets and such, or to just "hang out." Weed was smoked, wine was drunk and in general a very good time had by all in colorful costumes, flowers in the hair, flutes and guitars playing. A colorful scene of what seemed a youthful exuberance, and attracted Allen Ginsberg back, and Gary Snyder, and others who had been thought of as the Beat Generation. And also a former Harvard professor who had seemingly discovered anew a "miracle drug": Lysergic Acid, or LSD.

Dr. Timothy Leary had been making headlines for the past couple of years after having been fired, along with his associate Richard Alpert, from Harvard University. They had headed a research program named the Harvard Psilocybin Project from 1960–1962. There were allegations that they used undergraduate students to test the drug experimentally after being told by the university to only use graduate students. In addition, both Leary and Alpert took the drug along with their subjects. In 1963, both were fired. Alpert went on to become Ram Dass, a spiritual teacher with a large following. Leary promoted LSD publicly and acquired a 64-room mansion in Millbrook, New York. There, having conducted LSD sessions, the group became The Castalia Foundation, named after a colony in Hermann Hesse's *The Glass Bead Game*, a favored book of the '60s. All was centered on finding and cultivating the divine within each of us. The reputation of Leary and Millbrook grew and attracted the likes of Tom Wolfe, a New York writer, who wrote *The Electric Kool-Aid Acid Test* about Leary and his LSD doctrine. Always, there was the claim that he was only interested in research and not in using it merely for recreation. Over and over, Leary was described as "advocating the exploration of the therapeutic potential of psychedelic drugs under controlled conditions" – this phrase being used in numerous books and articles about him.

It is hard to reconcile this phrase with his initial appearance in the Haight-Ashbury counterculture on January 6, 1966, at what has become known as "The Human Be-In."

A few months earlier, on October 6, 1966, a California law making LSD illegal had been put into effect. Dr. Leary stepped onto the stage constructed in the Golden Gate Park Polo Field, and, to a crowd estimated later to be between 20,000–30,000 people, mostly from their early teens to twenties, all in the midst of a break from the values of their parents and the culture of the 1950s, proclaimed over the microphone "Turn On, Tune In, Drop Out!" Controlled conditions be damned, as Owsley Stanley had provided massive amounts of his White Lightning LSD to the crowd. And as Jefferson Airplane, The Grateful Dead and

Big Brother and the Holding Company blasted out their rock, many did, indeed, turn on and tune in – the drop out had occurred shortly before.

Dionysus had certainly landed in this San Francisco counterculture of the late '60s, but the indiscriminate use of what had been, more than 2,000 years ago, ritual use of a substance in a controlled environment turned into something much darker and more dangerous.

All this time, my then husband, a photojournalist of some standing, was documenting the scene. A friend of mine, a woman who wrote a column for the local San Francisco paper, was married to an obstetrician/gynecologist who volunteered at the Free Clinic in the Haight as he was aware of the less than hygienic conditions that many of the kids were living in there. One evening, called out to a young girl who was delivering her baby at home, he facilitated a safe and successful delivery. After a long siege, he accepted a glass of water from one of the group living there, and shortly afterwards began to hallucinate badly. Someone had the presence of mind to call an ambulance and he got to an ER safely. However, this event precipitated a complete schizophrenic breakdown, resulting in hospitalization and a year-long recovery. One of those kids had "gifted" the water with a couple of tabs of acid.

Over the course of that year or so, I accompanied my husband to various places in the Haight: the old Victorian house where The Grateful Dead lived, Michael Bowen's flat and others. But with two small children at home in Berkeley, I made it a rule never to eat or drink anything offered to me. In general, I saw groups of earnest, often very talented young people have fun, consciously breaking with the values of a generation before whose values they despised. But with all the Dionysian trappings, I saw very little of a spiritual nature or the introspection that LSD was touted as providing. Instead, I saw chaos.

It was several years later, studying at the C.G. Jung Institute for Analytical Psychology in Zurich, that I realized what had happened in the Haight. What had started out as a rejection of the mores of a generation, a discontent, had turned, with the help of drugs, into an out-of-balance situation that could only result in chaos. This was because the very Apollonian outlook of America in the 1950s had been so entrenched and fixated on a materialistic culture that any challenge would be out of balance. The Dionysian always seems to have its way in, upsetting the culture, rather than coming into equilibrium with it. Stability can only come through a balancing. The brief Dionysian fling of the 1920s burned itself out in bathtub gin, and, in the later 1960s, not only hallucinogens but also widespread use of heroin and cocaine tipped the scales.

However, if the Dionysian implies abandon, intoxication and creativity, it can come with death and chaos too, which are also part of the Dionysian.

And what of Dionysus in the twenty-first century? Early indications are that He is appearing in a slightly different form in this era. More like the Ancient Greek Festival held yearly in Athens from the fifth century AD: the Great Dionysia as the now modern annual Burning Man festival.

13 Burning Man

Dionysia into the twenty-first century

The Summer of Love celebrated its fiftieth anniversary in 2017; most of the Flower Children were then in their sixties and seventies and much nostalgia was in the air. Where had it all gone? The celebrations, the art, the rebellion in the very air of those brief few years?

Actually, just about twenty years after that peak in 1967, once again some friends put together what began as a small summer solstice celebration on a San Francisco beach. It grew and grew into the twenty-first century, when it became a phenomenon that has attracted global attention. And it continues to take place, once a year, in the week prior to Labor Day in a unique setting: the Black Rock desert of Northern Nevada. What is happening there seems to be the latest appearance of Dionysus. Ginette Paris has written of festivals that "there has to be a group emotion, a group enthusiasm for Dionysus" (Paris 1980 p. 22).

And that certainly seems to be taking place at what has become known as "Burning Man."

In the late decades BCE, a phenomenon known as the Greater Dionysia began in Athens, Greece. Prior to that, all over Greece there were the "lesser" Dionysia, which were small festivals held in rural locations in honor of the god. But it was in the Greater Dionysia in Athens that the festival burgeoned into something more:

> With this feast the Dionysian festive period in Athens, which opened with the beginning of winter, reached over to…March. Thus it exceeded in scope, not only the Dionysos cult of Delphi, which remained essentially a woman's cult but all other Dionysos cults of the same style.
>
> (Kerényi 1976 p. 316)

It opened with a grand parade in which an effigy of Dionysus led the way. In earlier centuries there was an enormous phallus instead of a statue of the god, representing Him and his archetypal power. The great tragedies of the times, written by Aeschylus, Euripides and other greats, were presented (even debuted) at this time, among all the revelry and celebration. "[T]ragedy reveals a conflictually constituted world defined by ambiguity, duplicity, uncertainty and unknowability, a world that cannot be rendered rationally fully intelligible…" (Critchley 2019 p. 87): in other words, tragedy is thoroughly Dionysian.

What makes Burning Man comparable to this ancient feast is that it, too, is the scene of extraordinary creative artistic activity and continuous celebration, often accompanied by the drinking of wine (and perhaps the taking of a few other substances, now that recreational marijuana is legal in Nevada). It was in the ancient festival that drama was born – the tragedies often associated with Attic Greece, that developed out of Dionysian ritual. And in its most modern form, in Nevada, art, architecture, sculpture and theater dominate the week-long celebration. And comedy, which might be said to be the latest manifestation of what is thought to be its birthplace in another Dionysia, the Athenian festival of Lenaia.

Dominating the seemingly barren landscape of the Black Rock Desert event is the Burning Man, a 40-foot sculpture of wooden lattice work which is set on a platform. Shafts of colored neon light illuminate it. It is filled with explosives which have carefully been orchestrated to go off in sequence, setting it afire at the conclusion of the festival – hence the name "Burning Man": another link to the Dionysian, where there was always a sacrifice, albeit an animal sacrifice. One story has it that at one of the Burning Man events, possibly in 2004/2005, the Hollywood actor Susan Sarandon threw some of the ashes of her deceased friend, Dr. Timothy Leary, into the blaze. True or not, it seems fitting that one of the most well-known people of the sixties should join in the modern Dionysia. Unfortunately, there has also been one case of self-immolation, too. Now the security at the site has been tightened to prevent further such happenings.

Many of those who come to Burning Man do so in costume harking back to the Haight-Ashbury days when brightly colored clothing, unique outfits and ethnic dress were worn by a large segment of the population However, at Burning Man anyone can wear whatever they want – or wear nothing at all.

> … for women…hot pants or body/bathing suit, ornate headdress, angel wings, feathers, bead, or even face coverings, leaving the breasts out or covered with just swatches of fabric. Looks awesome, but it is not exactly an aesthetic that lends itself to a variety of body types…But regardless of gender identity what I saw and felt

was that standard Playa outfits were mired in and sometimes even exaggerated, the sexualized objectification of women's bodies in mainstream media.

(Mukhopadhyay 2017)

The same atmosphere prevails (or did) as in the Haight: what one writer has described as Dionysian Anarchism – a distinct difference in philosophy, behavior and thought (Pendell 2008). Recently, however, there are indications of change, or perhaps evolution – although on the level of mythological thinking it might be said to be changing from a basically Dionysian model to an Apollonian one. This has much to do with the fact that besides being expensive, as it always has been – with large entry fees and requirements for water, food and shelter being supplied for a week by each participant for their personal needs in this very dry, hot and windy environment where only coffee and ice are sold – it is now attracting many millionaires, and even billionaires. They are, of course, bringing in (or having constructed by others) luxurious shelters for themselves, along with private chefs and personal attendants. A lot of business and networking is going on there between them now. The central, post-Hippie environment is at the center still, but this fringe group is having its impact on the whole. One writer states that the escalating ticket prices are "making it perhaps the consummate capitalistic counter cultural happening of our time" (Steinhauer 2018). And those prices are apt to go up, as recently (2019) the Bureau of Land Management of Nevada has requested that more security be provided for the event as it looks to increase capacity to 100,000 participants, citing it as a possible terrorist target. This will of course cost the Burning Man presenters even more money, and ticket prices will go up accordingly.

Some of these changes don't seem to coincide with the ten principles of the original founders: gifting, radical self-expression, decommodification and leaving no trace. Some are beginning to wonder if it has become a playground for the (mostly white) wealthy. Brian Doherty wrote, in the early days of Burning Man,

Uniting every divergent tendency, spirituality and attitude, is a sense that every life is missing something of a spark of creativity, a chance for self-expression, some freedom from judgement and cold personal relations that one must travel far off the grid to find.

(Doherty 2004).

There appears to be increasingly less of the above to be found in Burning Man as the rich set up their own version of paradise, one funded by wealth, and often mainly as spectators. And Dionysus? Perhaps he makes the occasional visit, especially in the artwork constructed and created at Black Rock, but somehow it no longer seems his territory.

References

Critchley, Simon (2019) *Tragedy, The Greeks and Us*. New York: Pantheon Press.

Doherty, Brian (2004) *This is Burning Man: The Rise of a New American Underground*. Self published.

Kerényi, Carl (1976) *Dionysos: Archetypal Image of Indestructible Life*. New Jersey: Princeton.

Mukhopadhyay, Samhita (2017) Burning Man promises to disrupt the modern world. And yet the gender dynamics are all too familiar. *Mic Network*. online

Paris, Ginette (1980) *Pagan Grace*. Dallas, Texas: Spring Publications.

Pendell, Dale (2008) Thoughts on Burning Man, the Green Man, and Dionysian Anarchism with Four Proposals. *Entheogen Review*, 17(1):10–14.

Steinhauer, Jillian (2018) The Vanishing Idealism of Burning Man. *New Republic Magazine Online*. August 22.

14 The return of Dionysus

What is being experienced all over the world in the twenty-first century is the eruption of the repressed Dionysian impulses and way of life. Repression at some point bursts onto the scene in a disruptive manner, stirring up what any given society has been living as their norm: "Dionysus arises in an age too dominated by the sun god Apollo's rationality" (Rowland 2017 p. 29). This is most recently to be seen in the Women's Movement of the last five years and in the LGBTQ movement, both of which have come up noisily all over the world. And Black Lives Matter, too, drawing our attention to what has been going on in terms of prejudice against African Americans since the Civil War (or the "War of Northern Aggression," as some Southerners still insist on calling it). New generations are striking out to be able to live their lives more true to themselves in a society which, for the past 2,000 years or so, has rejected them for various reasons, from religious beliefs to political affiliation – and generally out of fear.

More than 2,000 years ago women lived in equality, and from earliest times gender was seen and known to be fluid and respected not only among the gods, but among humans too. Dionysus had both male and female within him. And he was perhaps the only god in the Greek pantheon who showed and experienced what we now call bisexuality and transgender.

The myths of Dionysus are many, possibly due to the fact that his worship did not originate in Greece, but most probably came from the Middle East. He was said to be born male, but other stories have him brought up as female. In even later stories he became both male *and* female, suggesting once again his non-Greek origins, as most Greek deities are firmly male or female – no gender switching, or allusions to such.

Possibly it was this very aspect of Dionysus that made the Greeks of the time ambivalent in their feelings toward this God, who did not fit comfortably into either identity and who, further was considered a God of Women.

For the Greeks, Dionysus was an uncomfortable god in many ways because of his fluid gender. The Early Attic Greek society was already very much a patriarchal society where, as previously stated, women had very definite roles to fulfill as wives and mothers. Dionysus and his maenads were accepted nervously. He was one of the few gods in the Greek pantheon who was a friend to women, and we find no stories of him raping or seducing human women as many of the other male gods were said to do. The one human woman with whom he was identified was Ariadne, a Cretan princess who had given the secret of the labyrinth of the Minotaur to Theseus, a captive prince, who slew the Minotaur and was thus able to escape. He left Ariadne behind on an island when he returned home. It was here that Dionysus found her, took her as his mate and pled to the other gods to make her divine. Which they did.

The Greeks were more comfortable with the occasional pairing of male gods with one another, as long as both remained firmly male: Apollo and Adonis, Heracles and Hylas, Hermes and Crocus; and, on a human level, the love of Alexander the Great for his friend Hephaestion, whose death caused Alexander the most terrible grief that it was said by some to have eventually lead to his own death. There are many such stories of male love throughout European mythology. Homophobia was not practiced as such unless a female element crept into it. And this seems to be true of other European myths also, including Norse and Celtic.

However, this was very different in some Asian cultures. In Chinese myth, homosexuality was overt, even having a deity especially for those involved, in the figure of To Err Shan.

As for many Eastern religions, including Shinto and Buddhism, same-sex love, male or female, are not categorized or differentiated to any extent, except in so far as all sexual activity is seen as something of a hindrance to spiritual goals and attainment.

And in our own Native American cultures of North and South America, we see perhaps an even greater acceptance of sexual and gender differences, starting with the early Mayan and Aztec cultures and going to the cultures in place when North America and Mexico were conquered by the Europeans. Vasco Nunez de Balboa was said to have discovered a group of two spirit chiefs, both genders within them, in 1595 in Panama and he ordered them to be thrown to his dogs to be torn apart and eaten alive. Homosexual persecution had been going on for several hundred years in Europe by then.

In Inuit myth, the first humans created were both male, one of them converting to female to bear children. Sedna, one of their major goddesses, is attended by what they term "two spirited" or dual-sexed shamans, considered as spiritual beings. And Sedna herself is often depicted with a female partner.

In a recent article by a Jungian analyst, the author had this to say about a book she had reviewed a couple of years ago:

> The author wrote that Native Americans of long ago had many names for each of the genders they perceived – it is obvious they were far ahead of us….in addition to male and female genders, there are two more genders to consider. Few societies in the world have had the wisdom to recognize the worthiness and contributions of people who live beyond the binary gender labels. Now our own society, or at least part of it, is slowly coming to accept the reality that there are men and women who are of mixed gender or of opposite gender or fluid gender.
>
> (Frantz 2017 p. 271)

However, still there is a terrible prejudice, widespread at least in the United States, against anyone who presents beyond the "norm" we have accepted of being either male or female, the enforced binaries of largely the Western world. This seems to be generated by fear and supported by fundamentalist religions and very conservative political stances in their demonization of any other choice. Recently, we're seeing a minor break up of the rules as the women fight for equality in pay and stature and the LGBTQ folks in public life come "out of the closet," as the old saying goes. All these rights that are now being fought for are simply a demand for equality and balance in a society which, especially since industrialization, has become completely one-sided. As white people we have behind us centuries of ancestors who did not conform gender-wise, and it is that we must partially draw upon to right all the terrible wrongs that have gone on for centuries and continue to go on for women, people with gender differences and people of color.

References

Siobhan (May 6, 2017) *Let's Talk About Dionysus Gender Queer God of Parties and Pride.* Autostraddle.com: www.autostraddle.com/lets-talk-about-dionysus-genderqueer-god-of-partying-and-pride-379653/.
Frantz, Gilda (2017) Gender Diversity. *Psychological Perspectives, A Quarterly Journal of Jungian Thought*, 60(3). C.G. Jung Institute of Los Angeles.
LGBT Themes in Mythology. Wikipedia.com.
Rowland, Susan (2017) *Remembering Dionysus*. Oxford/London: Routledge.

15 Apollo/Dionysus, masculine/ feminine, yin/yang

Throughout this work on Dionysus and women and gender, these bin-aries come up. It seems inevitable, given that Western society dominates most of our thinking and has for more than 2,000 years. Some of the greatest minds of our era have been, and continue to be, dominated by the split into masculine/feminine or whatever we choose to label that dichotomy. This seems to stem, in so far as I've been able to research it, from the Indo-European cultures that were the beginning of our civilization. These were warrior peoples, patriarchal not only domes-tically, but in their mythologies too, which had few (if any) goddesses, but many gods: mighty warriors, heroic in battle and from whom their human rulers claimed descent. Women in these societies were seen to have value only as mothers and wives to these warriors. Once in a while we hear of a special woman who acted as a warrior and rode into battle, but it is her masculine qualities that are praised.

On a largely unconscious level, these ideas and modes of being still predominate, and lately, with the rise of demands for equality from various segments of society, the powers that be, along with their archaic thinking, are feeling threatened and trying to hold on even tighter.

C.G. Jung was one such thinker, who

> liked to try to stabilize his notion of gender by offering an uncon-scious feminine archetype in man that he called "anima" and a com-plimentary masculine figure in a woman he named the "animus". For Jung fidelity to his organic and creative idea of the psyche means that gender is fluid.
>
> (Rowland 2017 p. 4)

Yet some of the old stereotyping of gender affected his thinking too. The 'anima' in a man is often the fount of creativity and feeling. However, too often the animus in a woman is presented in a disagreeable light.

It is a term "used by Jungians to explain some of the more aggressive or assertive characteristics in a woman" (Anthony 2017 p. 216). Even Jung often regarded this contrasexual in women with fear and loathing: "A woman should constantly control the animus by undertaking some intellectual work, many a woman has been drawn to disaster by her animus"; and, "If a woman dreams of a superior role she wishes to assume in the world, its best to advise her to write an essay or an article about her wishes" (Samuels 1986 p. 216).

For those of us who hope for change, we must become aware of how the old, patriarchal beliefs still permeate every subject we study and everything we read, from the most intellectual right through into the more intuitive realms.

A case in point was brought forcefully in the 1990s via the subject of archaeology. Marija Gimbutas, an archaeologist, brought to public attention the fact that her field of study was, and is still, predominately worked by men. She spoke to the fact that on many of the digs in which she participated as a professional, her findings were often ignored or downplayed, particularly the finding of those female figures which were profusely scattered but often dismissed as merely "fertility" figures. She said that "Communities of priestesses and women's councils which must have existed for millennia in Old Europe and in Crete, persisted into the patriarchal era, but only in religious rituals." When she published her book, *Goddesses and Gods of Old Europe*, she had to convince her publisher to put "Goddesses" first, not "Gods."

Sir Arthur Evans, famed for his discovery and uncovering of Minoan Crete, seeing the multiple frescoes of women, mainly in lines as though going someplace, dismissed them as pretty women filling the court of King Minos. On closer observation, to even the untrained eye it becomes obvious they are bringing offerings to their goddess, she of the snake-entwined arms.

And in all the sciences, the role women have played has gone shamefully unreported. In 1962, Wilkins, Crick and Watson were awarded the Nobel Prize for their discovery and development of DNA. However, a chemist name Rosalind Franklin had worked on X-ray distraction images of DNA years earlier: "Her data were critical to [Wilkins,] Crick and Watson's work" (Lee 2013). Until recently, she received no recognition for her part in the discovery and was not included in the Nobel Prize as she had died in 1958 of ovarian cancer (Nobels are never given posthumously). But there was no mention of her, in any case.

There are many women in the sciences who have experienced similar invisibility when it has come to their accomplishments, Another woman, Jocelyn Bell Burnell, discovered pulsars as a graduate student in 1961. However, the Nobel Prize for that discovery went to her school supervisor. And it would be naïve to think that this is not still going on, though with the latest Women's Movement, which is bringing to light the deliberate exclusion of women in the continuing patriarchal thought in the world, one hopes it will soon come to an end.

The above examples, only a few within a multitude, make it imperative that in all we read and research, we must always take into account the society and its prejudices in which the writer was raised and lived. What we need desperately is balance; not masculine/feminine, Apollonian/ Dionysian, yin/yang as warring opposites, but as complementaries, each needing the balance of the other. This is not impossible to attain if the current patriarchic thinking succumbs and gives credit to the feminine in the form of equality and not continuing to impose the archaic, 2,000-year-old conceptions of gender on our modern culture. Three social movements – the Women's Movement, the LGBTQ community and Black Lives Matter – are fighting the good fight now, and all our lives will be richer if equality and balance come out winners.

In this work, I've mentioned two women whose work seems to point us in the direction needed to bring all this about – Clarissa Pinkola-Estes and Jean Houston – and I'm sure there are many others out there too, but these are the ones who have come to my personal attention as possible guides for the future.

Pinkola-Estes comes to mind through her usage of story and myth in forming one's life. Personally, this has been a very important part of my journey. At thirty-one years old, and just having my first child, a baby girl whom I adopted at birth, I went into Jungian analysis. What triggered my going was that at that point in my life I was becoming aware that patterns could be, and often were, repeated in relationships. My relationship with my mother was a troubling one for me, a case of "smother" love in which I realized I was a willing participant. This gave me a great deal of guilt at how dependent I was on her still, even at my age. When my husband's mother gave us a lovely home in Berkeley, California, we moved without hesitation from our rented Victorian flat in San Francisco. My daughter was just three months old and I wanted her to have what I never had as a young child: her own room. However, I realized that I had no friends in Berkeley, as they all lived in San Francisco. And my husband's profession as a photojournalist meant that he would be spending most of his working hours across the Bay in San Francisco. So I made it a proviso of the move that my mother would live with us. She was widowed and had a small fixed income, but mostly I wanted her with us so I had another adult person in the house for company. I didn't want nor ask her advice about raising my baby as

I didn't feel I had been raised well, and didn't want my relationship with my daughter to run the same course as mine had with my mother.

I was aware I needed help with my need for her presence, which contained a large degree of toxicity for me. So analysis seemed a good place to work it out and look for some answers. And Jungian analysis seemed appropriate, since it had been only three years since I had read *Memories, Dreams, Reflections* in Rio de Janeiro and had found a kindred spirit in Jung and his view of the world.

In my first session with Dr. John N.K. Langton, he sat across from me, notebook and pen in his hands and said "Tell me the story of your life." BAM! I was shocked and tried to think how I could speak of all those thirty-one years in fifty minutes. But I began, and all the important events seemed to pour out into a mini-biography. When I finished, Dr. Langton got up, walked to a small window in the office, briefly looked out, ran one hand through his hair and said, in an excited tone, "Don't you see? You've been living the myth of Persephone!"

I had always loved mythology since my earliest years, especially Greek mythology, so the story was known to me. I caught his excitement, and in that moment I saw the story between my mother and myself stretch back 2000 years, through that story. In that moment, my guilt dropped away and never returned. He had presented me with the story which provided the reasons for my feelings, plus a way to get beyond my role in that myth. I had been the daughter, Persephone, but now was the mother, Demeter. And I realized it and could work with that, as now, in my older age, I am the Old Woman, Hecate, as well, and am working through that.

Jean Houston's way is slightly different, but just as effective for many. Her emphasis these days is on small groups (limited to twenty people) coming together in her home in Ashland, Oregon, for three or four days for intense seminars with her. These are highly experiential and geared toward creating greater personal awareness for the empowerment of one's self, with an emphasis on one's creativity and the addition of practical steps to attain goals. In addition, for three months afterwards, there are monthly teleconference calls with Houston. A personal mentoring. An emphasis on one's own story and changing it for the betterment of one's life and work.

Of course there are other women out there all working in their unique ways toward the re-education of women and our thinking about what is feminine to help women realize the full power they have, which has been taken away for the last 2,000 years.

Thinking of this brings to my mind the late science fiction master writer, Frank Herbert. He was the author of one of the most powerful, best-selling series of books that began with *Dune*. In it he writes of a fictional world in which thinking machines have taken over completely – a world which, when *Dune* was published in 1962, seemed quite fantastic. Now, in the twenty-first century, it seems eerily visionary of a time that is already beginning to become possible. In this book, there is a revolution against the machines: people destroy them and take back their civilization and power. Behind all this is a women's political/religious group called the Bene Gesserit. They undertake the re-education of the people and see to it that not only do they come into their own full intelligence, as brilliant as any of the artificial intelligences to whom they had once handed over their society, but that they have use of what the machines never had: the feeling and intuition of the human mind, heightened to full capacity.

In reading Dune once again, more than fifty years later, I think of wise women such as Clarissa Pinkola-Estes and Jean Houston and others and wonder if they are OUR Bene Gesserit and can save our humanity from the increasingly negative road we have taken, forgetful of our long past, and gear us up with realization of our full power and need at this time.

References

Anthony, Maggy (2017) *Salome's Embrace: The Jungian Women*. London: Routledge.

Lee, Jane J. (2013) Six Women Scientists Who Were Snubbed Due to Sexism. *National Geographic*. May 19: www.nationalgeographic.com/news/2013/5/130519-women-scientists-overlooked-dna-history-science/.

Rowland, Susan (2017) *Remembering Dionysus*. Oxford/London: Routledge.

Samuels, Andrew (1986) *Jung and Post-Jungians*. London: Routledge.

Index

For Product Safety Concerns and Information please contact our EU representative GPSR@taylorandfrancis.com Taylor & Francis Verlag GmbH, Kaufingerstraße 24, 80331 München, Germany

Printed and bound by CPI Group (UK) Ltd, Croydon, CR0 4YY

11/04/2025

01844010-0002